Power Systems

Electrical power has been the technological foundation of industrial societies for many years. Although the systems designed to provide and apply electrical energy have reached a high degree of maturity, unforeseen problems are constantly encountered, necessitating the design of more efficient and reliable systems based on novel technologies. The book series Power Systems is aimed at providing detailed, accurate and sound technical information about these new developments in electrical power engineering. It includes topics on power generation, storage and transmission as well as electrical machines. The monographs and advanced textbooks in this series address researchers, lecturers, industrial engineers and senior students in electrical engineering.

Power Systems is indexed in Scopus

More information about this series at http://www.springer.com/series/4622

Yuan Jiang · Qing Li

Vacuum Circuit Breaker for Aviation Variable Frequency Power System

Theory and Application of Arc in Electrical Apparatus

Science Press
Beijing

Springer

Yuan Jiang
School of Automation
and Electrical Engineering
University of Science
and Technology Beijing
Beijing, China

Qing Li
School of Automation
and Electrical Engineering
University of Science
and Technology Beijing
Beijing, China

ISSN 1612-1287 ISSN 1860-4676 (electronic)
Power Systems
ISBN 978-981-33-4780-9 ISBN 978-981-33-4781-6 (eBook)
https://doi.org/10.1007/978-981-33-4781-6

Jointly published with Science Press, China
The print edition is not for sale in China Mainland. Customers from China Mainland please order the print book from: Science Press.

This Springer imprint is published by the registered company Springer Nature Singapore Pte Ltd.
The registered company address is: 152 Beach Road, #21-01/04 Gateway East, Singapore 189721, Singapore

Preface

In more electric aircrafts, electricity is generated by a frequency-varying generator and the current frequency is in the range of intermediate frequency (IF) 360–800 Hz, which is called a variable frequency (VF) power supply system. As the frequency and current increase, the breaking process becomes difficult and hence new circuit breakers are required to ensure the safety of the aircraft. Vacuum circuit breakers have been widely used in civil power systems, and are potentially suitable for application in the power supply system for aircrafts. Current breaking in vacuum circuit breaker is a systematic theory to study the physical processes that occur between two contacts, related to many fields, such as electric, mechanic, materials, computational technology, and so on. The goal of the research is to improve the reliability of breaking process and the life of the vacuum circuit breaker under the premise of meeting the economic and safety benefits.

This book focuses on the vacuum circuit breaker in variable intermediate frequency power supply system, mainly involving characteristics of current interruption and breakdown, as well as arc theory. In this book, the relevant theories, devices, tests, simulation, and experimental methods of the vacuum circuit breaker in variable intermediate frequency conditions are introduced in detail. The effect of frequency on characteristics of vacuum arc and the mechanism of breaking and post-arc breakdown under different operating conditions are discussed.

Combined with the practical application, the effect of contact structure on the breaking characteristics, such as arc mode, arc pinch, and magnetic field in a vacuum circuit breaker are analyzed, and the test technology of VF vacuum circuit breaker is proposed. The post-arc breakdown problems, involved in the voltage spike, marginal effect on field emission, and sputtering of macroscopic particles, are deeply discussed by the aspects of basic theory, research methods, and simulation. The whole book (or part of the chapters) can be used as a textbook or reference for graduate and senior undergraduates.

I would like to thank my wife, Ms. Qian Zhang, for her contribution to the translation of this book and her support for my scientific research.

The book is supported by Guangdong Basic and Applied Basic Research Foundation (2020A1515110725).

Due to the limited level of the author, there are still some shortcomings in the book, please correct.

Beijing, China Dr. Yuan Jiang

The original version of the book was revised: Funding information has been updated. The correction to this book is available at https://doi.org/10.1007/978-981-33-4781-6_6

Contents

Chapter 1
Introduction

1.1 More/All-Electric Aircraft and Aviation Power Systems

1.1.1 More/All-Electric Aircraft

As a strategic industry, aviation industry is an important symbol of science and technology, industrial base, and comprehensive strength, and has been highly valued and given priority to development by all countries worldwide. With the application of new technologies such as electromechanical actuators and power by wire (PBW), the onboard load capacity has been greatly improved. While the power consumption has been increased and the power supply quality and reliability have been improved, the existing power system is unable to meet the power demand, thus promoting the development of large capacity aviation power supply system. Facing the future, the secondary power system of aircraft will gradually be unified into the electric power system, from the traditional multi-energy system such as mechanical energy, hydraulic energy, and pneumatic energy. That is named more/all-electric aircraft. Electrification has the advantages of low fuel consumption, high reliability, high maintainability, and high energy conversion efficiency, as well as is an important way to support green aviation and the development trend of aviation industry. Since 1990, the Department of Defense and NASA in the USA have been conducting demonstrations and research on more/all-electric aircraft. By 2012, they have completed the improvement of the generation form and power level of the DC power supply system for F-22 and F-35 aircraft in three stages. Currently, the MTA program is being promoted to achieve Megawatt generation power of Tactical Aircraft [1, 2]. Since 2013, "Three Musketeers" in China have made their first successful flights, among which the large conveyer Y-20 and the large passenger aircraft C919 have adopted more electric technology.

Y. Jiang and Q. Li, *Vacuum Circuit Breaker for Aviation Variable Frequency Power System*, Power Systems, https://doi.org/10.1007/978-981-33-4781-6_1

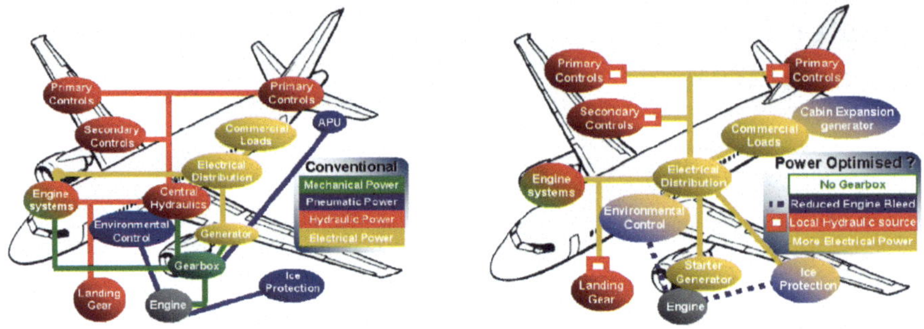

(a) Schematic of conventional power distribution **(b) A potential optimized architecture**

Fig. 1.1 Energy systems in conventional and more electric aircraft

In the energy system of traditional aircraft and more electric aircraft, as shown in Fig. 1.1, the traditional aircraft secondary energies include mechanical power, pneumatic power, hard power, and electrical power. The main secondary energy of more electric aircraft is electric energy and PBW technology is adopted, and hydraulic energy is used in local parts [3, 4].

1.1.2 Aviation Power System

1.1.2.1 Aviation AC Power System

The AC power systems currently used in more electric aircraft, including constant speed and constant frequency (CSCF), variable speed and constant frequency (VSCF), and variable speed and frequency conversion (VSVF), are shown in Fig. 1.2.

Most of the traditional AC power generation systems in aviation use CSCF power supply scheme with an intermediate frequency (IF) of 400 Hz. The aviation generator is connected to the engine through the constant speed mechanical transmission device to make the power frequency stable. However, the device is large and inefficient. The CSCF power system has the following problems: large volume of mechanical structure, low efficiency, low reliability, large amount of maintenance tasks, etc. It is difficult to increase capacity further. In the world, the maximum capacity developed is 150 kVA, and the maximum installed capacity is 120 kVA for Boeing B777 aircraft. In China, the maximum capacity developed is 90 kVA and the maximum installed capacity is 60 kVA.

Compared with the CSCF power system, there are some advantages for the VSCF power system, such as no constant speed device, no high stress mechanical/hydraulic components, or easily worn parts. It overcomes the disadvantages of constant speed device, which has high reliability, good maintenance, and low life cycle cost. It is highly efficient at generating electricity, increasing efficiency by 10%. At the same

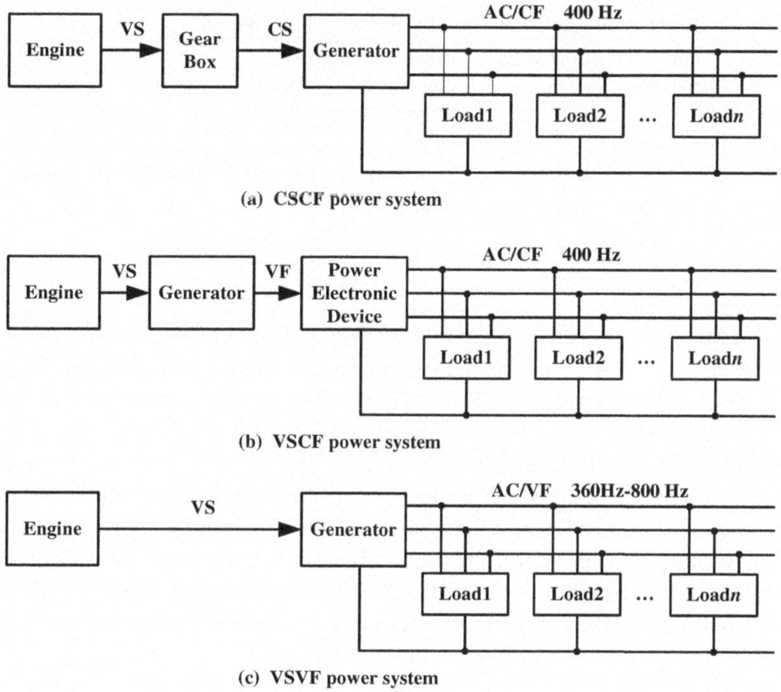

Fig. 1.2 AC power system for more electric aircraft

time, the power quality is better, and the voltage precision is higher, and the voltage transient is smaller. Unfortunately, there are also some new technical problems. For example, constant frequency is realized by power electronic devices. Due to the limited power, the capacity is generally 40 kVA, and the capacity on MD-90 aircraft is 60/75 kVA, which has reached the limit, and the temperature requirement of the installation environment is strict. There is a problem of compatibility with low power factor loads, such as inductive and capacitive loads. Therefore, the VSCF power system has been used as auxiliary power for a while in B777, and all of them have been replaced already.

The most advanced aviation variable frequency (VF) generator can connect the generator to the engine, making the frequency of the power supply system variable IF 360–800 Hz. The VF AC power supply system is composed of an alternator and controller, which has only one transformation process. The alternator is directly driven by the engine motor box, without a constant speed transmission mechanism or constant frequency power electronic device. The system has the characteristics of simple structure, lightweight, small volume, high power density, high reliability, low life cycle cost, good maintainability, and high energy conversion efficiency. However, there are also some technical challenges, such as for load adaptation, resulting in an increase in the volume and weight of electrical equipment,

Table 1.1 AC power system structures of typical more electric aircraft

Aircraft model	Main generator capacity (kVA)	Line voltage (V)	Main generator frequency (Hz)	APU capacity (kVA)	APU frequency (Hz)	Emergency power capacity (kVA)
A320ME	4 × 75	200	360–800	2 × 120	400	2 × 80
A380	4 × 150	200	360–800	2 × 120	400	70
B787	4 × 250	400	360–800	2 × 225	400	10

and the addition of controllers. According to the public information, the more electrification of large aircraft adopts the AC power generation with variable frequency scheme (the frequency range is 360–800 Hz). The typical representatives are the large airliners Airbus A380 and Boeing B787 [4]. The VSVF power capacity of A380 is 600 kVA, the line voltage is 200 V, and four 150 kVA alternators are driven by the main engine. Boeing B787 is closer to all-electric aircraft, with a total VSVF power capacity of 1.4 MVA and line voltage of 400 V. The main engine drives 4 × 250 kVA AC starters/generators, and the auxiliary power unit drives 2 × 225 kVA AC starters/generators. In addition, both the Y-20, a large transport aircraft developed in China, and the C919, a large passenger aircraft developed under the auspices of COMAC, have adopted VSVF power supply (Table 1.1).

1.1.2.2 Aviation HVDC Power System

With the increase of power consumption, power quality, and reliability, the traditional 28 V low-voltage DC power supply has been unable to meet the needs of modern aircraft, which promotes the development of the 270 V high-voltage DC (HVDC) power supply system.

The transmission lines of HVDC power system are light in weight, simple in structure, high in power density, high in energy conversion efficiency, and easy to realize uninterrupted power supply, which has become the first choice of power supply system for military aircraft. For example, F-22, F-35, and reconnaissance/attack helicopters RAH-66 are all equipped with HVDC power systems. The F-14A, S-3A, and anti-submarine aircraft P-3C are also partially powered by HVDC. For civil aircraft, although the main power supply of Boeing B787 is the VF alternator, 4 × 150 kW power supply with ±270 V is formed through the rectifier and other power electronic devices (Table 1.2).

Table 1.2 HVDC power system structure of typical more electric aircraft

Aircraft model	Main power capacity (kW)	APU capacity (kW)	Voltage level (V)	Short circuit current (kA)
F-22	2 × 65	22	270	2.5
F-35	2 × 125	200	270	5

1.1.3 The Technical Difficulties in the New Aviation Power System

With the development of new aviation power system, it brings technical problems. For example, when a fighter plane flies to a height of 20,000 m, the air pressure is only about 5.5% of the standard atmospheric pressure. The environment is humid and the temperature difference is large, so the requirements of insulation for electrical equipment are strict. Inductive load and distributed inductance of power supply transmission line will produce high overvoltage when current is switched off. Due to the limited space in the aircraft, the switches are required to be small in size, light in weight, safe in use, high in reliability, and easy to maintain. Air or inflatable circuit breaker was widely used in the field of aviation generators for protection. However, due to its limited breaking capacity, it cannot meet the requirements of high capacity of modern aviation power systems. It can be seen that there is an urgent need to design a new type of aviation switch apparatus to meet the application requirements and ensure the safety and reliability of power supply.

1.1.3.1 Protection of Aviation AC Power System

For the fault protection of aviation VSVF power system, there is no special protection electric apparatus at present, and air switch is still widely used. The working principle of air switch is to make the arc enter the extinguishing region from the contact area under the driving action of airflow and magnetic field. The arc is fully stretched and cut by the grid plate, and the arc extinguishing is taken advantage of by the near cathode effect and cooling effect [5]. As the aviation working environment has strict requirements on the size and weight of components, the arc interrupted the ability of "elongated" arc electrical apparatus is greatly limited by the size of arc extinguishing devices. Especially in the condition of VF 360–800 Hz, the current frequency increases by 7–16 times compared with the power frequency of 50 Hz, which makes the change rate of current di/dt faster. As the current increases, the arc extinguishing process becomes more difficult. Arc will cause the contact to melt, aggravating the electrical wear of the contact and destroying the insulation performance of arcing device, which will not only seriously affect the service life and reliability of air switch, but also limit the capacity increase of VSVF power system. It can be seen that the aviation apparatus are the key components in more electric aircraft. Therefore, it is urgent to develop new high-performance VF switches to meet the application requirements of future more/all-electric aircraft.

1.1.3.2 Protection of Aviation HVDC Power System

The development of aviation HVDC power system has important significance for improving the reliability, fault tolerance, power density, and flight performance in aircraft systems, which makes the importance of power supply and distribution system more prominent. DC contactors, circuit breakers, and power switches are the key components in the power distribution, management, and control systems to perform the functions of interrupting, power switching, and fault protection. Because there is no natural zero crossings, the difficulty of arc extinguishing increases.

1.2 Electrical Apparatus for Aviation Power System

1.2.1 Electrical Apparatus and Circuit Breakers

Electrical apparatus are electrical installations, equipment, and components used for circuit switching, circuit parameter transformation for control, adjustment, switching, detection, and protection of the circuit or electrical equipment. Circuit breakers are the most commonly used apparatus in power systems protection.

The circuit breaker is a switching device capable of closing, carrying, and breaking currents in normal and abnormal circuit conditions within a specified time. According to the use range, it is divided into high-voltage circuit breaker and low-voltage circuit breaker. The boundary of high and low voltage is fuzzy, and 1 kV is referred to as high-voltage electrical apparatus generally. Scientific and reasonable selection of circuit breakers is the key to ensure the safe and effective operation of the power system. It is necessary to avoid improper selection and unreasonable installation, which may not play the role of control or protection, and cause potential safety problems. The main technical parameters of the circuit breaker include rated voltage, rated current, breaking short circuit current, and so on. The rated voltage is related to the capacity and type of use. The same circuit breaker may have several rated working voltages. The rated insulation voltage is the maximum rated operating voltage of the circuit breaker, which can determine the electrical clearance and creepage distance. In no event, the maximum rated operating voltage should exceed the rated insulation voltage. The short circuit current of power system is usually ten to dozens of times the rated current. The short circuit current can be divided into three-phase short circuit, two-phase short circuit, and single-phase ground. The three-phase short circuit current is the largest, generally 20–35 times the rated current. The two-phase short circuit current is generally about 20 times the rated current. The single-phase ground current is the smallest, generally 8–15 times the rated current. The probability

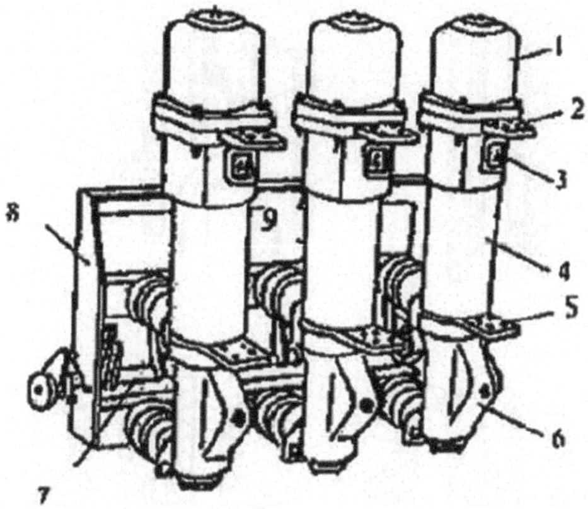

1. Upper cap, 2. Upper outlet, 3. Oil pointer, 4. Insulation cover tube, 5. Lower outlet, 6. Base, 7. Spindle, 8. Framework, 9. Trip spring

Fig. 1.3 The structure of less-oil circuit breaker

of three-phase short circuit current is the lowest and the probability of single-phase ground current is the highest.

According to the arc extinguishing medium, the medium voltage circuit breaker has less-oil, compressed air, vacuum, SF_6, and other forms. The less-oil circuit breaker (as shown in Fig. 1.3) refers to the circuit breaker that uses transformer oil or special circuit breaker oil as the insulation and arc extinguishing medium between contacts and uses solid insulation parts for ground insulation. The action and arcing chamber type of transformer oil are basically the same as that of more-oil circuit breaker, however, the oil amount is much less than that of more-oil circuit breaker. Less-oil circuit breaker is mainly developed in European countries, and the purpose is to save oil. Since the development of 220 kV low-oil circuit breakers in China in the 1960s, the switching current has developed from 16 to 40 kA. Less-oil circuit breaker is still dominant in 110–220 kV system in China and reaches 330 kV level. The less-oil circuit breaker for 10 kV has been replaced. However, worldwide, the less oil circuit breaker development process is close to the end. It has been gradually replaced by SF_6 and vacuum circuit breakers.

The compressed air circuit breaker uses high-pressure air to blow and extinguish the arc (in Fig. 1.4). The arc extinguishing chamber is mainly composed of nozzle. The arc is burned at the nozzle and the cooling effect is used to extinguish the arc. Therefore, it is an external energy type arc extinguishing device. The arc extinguishing ability mainly depends on the flow rate and airflow velocity at the nozzle. Increasing the working pressure of compressed air is the most effective measure to improve the breaking capacity of the compressed air circuit breaker. At the early stage, the working pressure was mostly 1–2 MPa, and now 3–4 MPa has been widely used, with individual products up to 5–6 MPa. Its disadvantages are complex structure, high processing, and assembly requirements, more non-ferrous metals, the higher price than less-oil circuit breakers, and the use of additional air

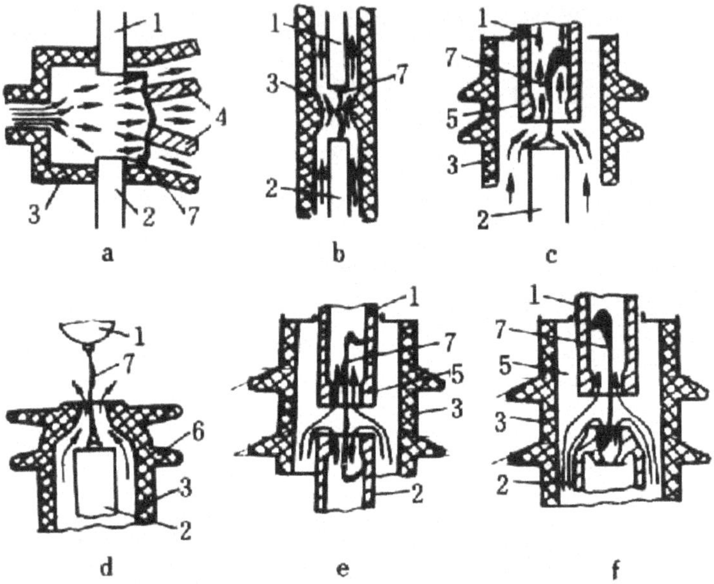

Fig. 1.4 The type of compressed air circuit breaker

compression devices. At present, it is used in high-voltage and ultra-high voltage levels, and in the medium-voltage gradually replaced by vacuum circuit breakers.

SF_6 has been used as an arc extinguishing medium for circuit breakers since the early 1950s. Due to the excellent characteristics, insulation and arc extinguishing performance of SF_6 are much better than that of compressed air circuit breakers and less-oil circuit breakers with the same breaks, and high pressure is not required. In the 1960s and 1970s, SF_6 circuit breakers (in Fig. 1.5) have been widely used in ultra-high voltage and large capacity power systems. In the early 1980s, SF_6 circuit breaker with a single break of 363 kV, a double break of 550 kV, and rated breaking current up to 100 kA has been developed. SF_6 is very stable at room temperature, however, if there is more water, hydrofluoric acid may be generated when the temperature is above 200 °C. To limit the production of toxic substances, moisture content in SF_6 must be limited. The consumption of SF_6 circuit breaker is always large, which is mainly used in voltage level above 110 kV. However, for SF_6 is a greenhouse gas, it is an inevitable trend to be replaced by other environment-friendly gases. At present,

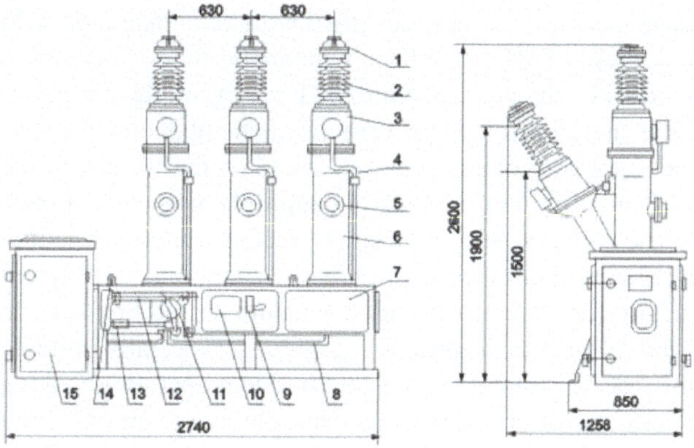

Fig. 1.5 The structure of SF_6 circuit breaker

it has been gradually replaced by vacuum circuit breakers at a medium voltage of 35 and 10 kV.

Vacuum circuit breaker (as shown in Fig. 1.6) is named because the arc extinguishing medium and the insulating medium are both vacuum. When dynamic contact moves, metal steam appears on the contact surface at high temperature, and arc is

Fig. 1.6 The structure of vacuum circuit breaker

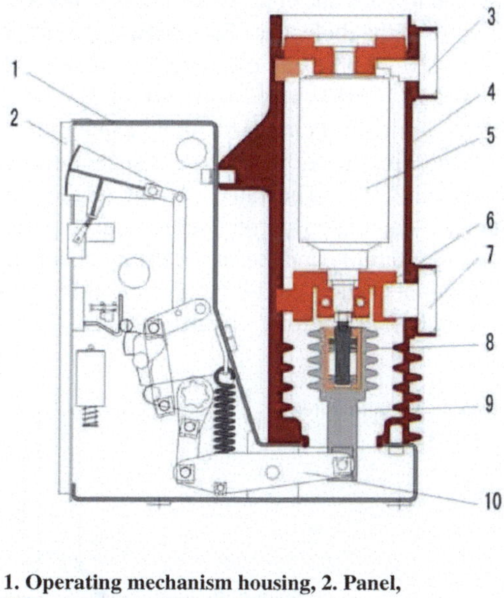

1. Operating mechanism housing, 2. Panel,
3. Upper outlet, 4. Insulation tube,
5. Vacuum Chamber, 6. Conductive clip,
7. Lower outlet, 8. Contact spring,
9. Insulating rod, 10. Transmission arm

generated between the contacts. As the contact is designed with axial magnetic field (AMF) or transverse magnetic field (TMF) structure, a magnetic field is generated when the current passes contact. For the contact with AMF structure, the arc maintains the diffusion state and the energy is low to prevent the anode phenomenon. For the contact with TMF structure, the arc rotates at a high speed on the surface of the contact to avoid excessive local ablation. The arc goes out naturally after current zero, and the strength of the dielectric between the contacts will recover quickly. Vacuum circuit breaker has the advantages of small volume, lightweight, suitable for frequent operation, and is widely used in distribution networks. Vacuum circuit breaker can be used for the protection and control of electrical equipment, especially suitable for the use place requiring oil-free, less maintenance, and frequent operation. Taking ZW8-12 series outdoor vacuum circuit breakers as an example, the main operating parameters are as follows: rated voltage AC 12 kV, rated trip current 630 and 1250 A, short circuit breaker current 12.5, 16, and 20 kA, rated peak withstand current 31.5, 40, and 50 kA.

For low voltage under 500 V, low-voltage circuit breaker, also known as air switch is used, with air as the arc extinguishing medium. When the switch is opened, an arc is generated between the separated contacts. The magnetic field generated by the arc current and induced by the arc extinguishing gate of the air switch attract each other, and the arc is pulled towards the arc extinguishing gate (as shown in Fig. 1.7). The principle of arc extinction is to cool the arc to reduce the thermal dissociation, and to strengthen the recombination and diffusion of charged particles by elongating the arc, and to blow the charged particles in the arc gap to quickly recover the insulation strength of the dielectric. Take ABB distribution network low-voltage circuit breaker CJX7-210-30-11 as an example, the main working parameters are as follows: rated voltage AC 380 V, rated trip current I_n 210 A, short circuit breaking capacity I_{cs} 35 kA, limit short circuit breaking capacity I_{cu} 50 kA.

According to "Low Voltage Distribution Design Specification GB50054-2011", in article 3.1.1, "The rated current of the electrical apparatus should not be less than the calculated current. Electrical apparatus should meet the requirements of dynamic stability and thermal stability under short circuit conditions. Apparatus used to disconnect short circuit current should be capable of switching on and breaking in short circuit conditions". In article 6.2.1, "For short circuit protection of distribution

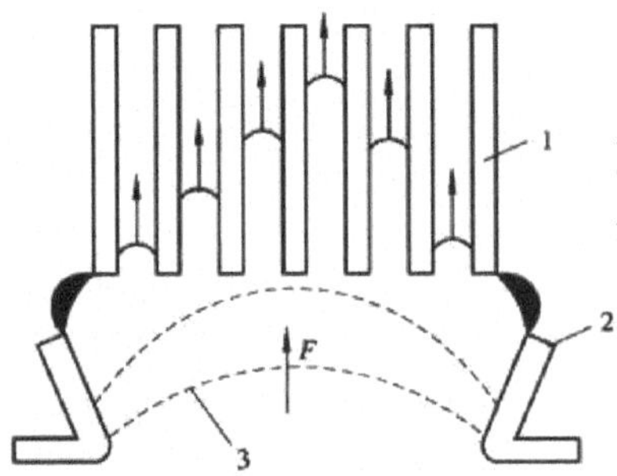

Fig. 1.7 The structure of arcing gate

lines, the power should be cut off before the short circuit current causes thermal and mechanical". In article 6.2.4, "When the short circuit protection device is a circuit breaker, the short circuit current at the end of the protected circuit should not be less than 1.3 times the setting current". The setting principle of low-voltage circuit breaker overcurrent protection is: When the electrical equipment is working normally and starting normally, the circuit breaker should not cut off the circuit. When the equipment is starting normally, the circuit breaker should not trip. And in case of fault, the protective apparatus should selectively cut off the circuit. The long delay type is mainly to protect overload. The setting current should be greater than or equal to the calculated load current, which can be determined as 1–1.1 times of the calculated load current, and not greater than 0.8–1 times of the long term allowable current. When selecting the setting value of short delay adjustable action current, the peak current of short-time load should be avoided, and the peak current should be calculated according to the starting current of the maximum motor. When selecting the setting value of the instantaneous characteristics of adjustable action current, the peak current should be avoided. The factor of the maximum full start current of a motor should be considered, and the reliability coefficient of 1.2 should also be considered. For the motor control system circuit breaker selection, the rated current selection is generally 1.5–2 times the rated current of the motor, in order to avoid the motor starting current. In summary, the rated current of the low-voltage breaker used in the distribution network is generally 1–2 times of circuit current. The price of low-voltage circuit breaker is related to the rated current and short circuit current breaking capacity.

1.2.2 Aviation Electrical Apparatus

Aviation electrical apparatus mainly includes remote control components, such as relays and contactors, and include protection components, such as fuses, circuit breakers. Aviation electrical apparatus are very important. Take the relay as an example: there were 45 relays in the early transport aircraft. By the 1950s, there were 90 relays in large transport aircraft. In the 1970s, 200 relays were used in large transport aircraft. However, in modern civil aircrafts such as B757 and MD-82, there are more than 1000 relays and contactors installed in various equipment. With such a large number, if one of them fails, the whole avionics system may fail, which is of great importance.

According to the structure, common aviation relays are divided into the electromagnetic relay, solid-state relay, hybrid relay, special relay (including polarization relay, reed relay, bimetal relay, thermal sensitive relay). The load current is generally limited within 25 A. For example, the KM series relays of LEACH can reach 50 A/28 V DC, while the KX series can reach 75 A/28 V DC. FCA-150 series of TYCO can also reach 50 A/28 V DC. There is no strict distinction between relays and contactors, and the basic structure of aviation contractors is similar to that of electromagnetic relays. The contact system of the contactor is mostly equipped with a special arc extinguishing device and buffer spring. In early aircraft sealed contactors, suction and double breakpoint bridge electric shock are used. In later contactors, balanced armature and balanced force electromagnetic systems are used, and the rated current can reach several hundred amps. For example, the rated capacity of the M707 series contactor manufactured by LEACH can reach 700 A/28 V DC, and the maximum capacity of the AC single-phase 200 V/400 Hz series contactor can reach 385 A. The 270 V DC contactor, produced in collaboration with LEACH and GIGAVAC, uses hydrogen as an arc extinguishing medium and has a rated breaking current of 100 A. TYCO also has contactors using the same arc extinguishing medium.

In the protection of aircraft power systems, fuse and circuit breaker are mainly used. The fuse serves to protect the circuit by burning out when the current is overloaded. There are many types and specifications of fuses used in aircraft, with hundreds of specifications selected by the International Organization for Standardization (ISO). Fuses are divided into fusible fuses, refractory fuses, and inertial fuses according to the function of the protection circuit. The minimum specification currently available is 25 mA. Fuses have been widely used on airplanes because of simple and reliable construction. However, fuses are disposable overcurrent protectors and have the potential to explode when melted down.

Aviation circuit breaker, also known as automatic protection switch and jump switch, refers to a high power switch device with the ability to repeatedly close, load, and automatic cut off normal and fault current, used in power system protection. In 1885, the earliest circuit breaker appeared in the world, which was a combination switch with a knife opening and an overcurrent trip. In 1905, air circuit breakers with free trip devices appeared. From 1930 to now, with the discovery of arc principle and invention of arc extinguishing device, the present circuit breaker gradually formed. In the late 1950s, new electronic components led to the introduction of electronic trippers.

1.3 Research Status of Circuit Breaker in Aviation Power System

1.3.1 Research Status of Aviation Circuit Breakers

The U.S. military is already quite sophisticated in aviation circuit breakers. According to the AD report in 1990, the military intelligence report no. ADA369412 showed that the remote-controlled circuit breaker combined with thermal free tripping and traditional tripping used by the Apache AH-64A had been operating steadily for 15 years, with only 1% experienced problems. In a technical report entitled "Solutions and Cost Performance of 270 V DC Contactors" published by Avionics Magazine, the 270 V DC contactors developed by LEACH and GIGAVAC can withstand 10 times of overcurrent. The contactors can be used in the aerospace field, and are light and affordable. EATON has successfully developed a 270 V DC remote circuit breaker with rated current 50 A.

Aviation circuit breaker products in China mainly include ZKC and ZKP, both of which are mainly used in transport aircraft. ZKC protection switch is composed of a common switch and bimetallic power off device. The switch is a non-free trip type, which can be forced on in a very urgent state. The ZKP switch is different from ZKC switch. Once the overcurrent fault occurs and the bimetallic strip acts, it cannot be connected in a short time. The circuit can only be connected again after the bimetallic strip is completely cooled down. Other aviation circuit breaker models are DBA, DBB, DBC, DBF, DBG, DBH, DBJ, DDB, SDB, a total of more than 50 specifications. The circuit breaker current covers 1–300 A, and all circuit breaker products are subject to GJB5885-2006 or GJB1932-1994 aviation circuit breaker standard. LGA circuit breaker, which is widely used in Europe and America, is a small pushbutton circuit breaker produced by CONVENTRY, with a rating of 13 different specifications from 0.5 to 40 A, and it is used in 28 V DC and AC single-phase 200 V/400 Hz system.

With the rapid development of power electronics technology, it is possible to bear the contact wear by current through power electronics devices and increase the reliability and life of circuit breakers. Since the on-state voltage of power electronic devices is lower than that of arc, the application in the aviation field has a broad prospect. Solid-state power controller (SSPC) is electronic aviation switching equipment. It is an intelligent switching device composed of semiconductor device, which is used to switch on and off circuits and report status information. It adopts the semiconductor device, which can realize the no-arc breaking. For there is no moving device, thus there is no mechanical wear and it is suitable for aviation applications. At present, companies studying SSPC are mainly DDC and LEACH. For DDC, rated current of 28 V DC SSPC is up to 25 A. For LEACH, rated current of 28 V DC SSPC up to 150 A and rated current of AC 200 V/400 Hz SSPC is up to 25 A. The main disadvantages of SSPC are that the on-state voltage drop is high, and the switching capacity is far less than that of the switch with contact, so it is difficult to replace the circuit breaker. In the late 1970s, the hybrid switching technology combining mechanical contact switch with power electronic devices was rising. Through the effective coordination of the timing sequence between the two, the arc is extinguished successfully and the intelligent control of the switch is realized. The hybrid circuit breaker in aviation is still subject to power, volume, weight, and other adverse factors.

With the maturity of VF power generation technology, the capacity of aviation power systems has been greatly increased. A large increase in the power generation will make it more difficult to break the fault current. If the circuit breaker is used in the VF system of more/all-electric aircraft, the problem of arc breaking in high frequency and high current needs to be solved. The arc extinguishing ways of conventional aviation circuit breaker include simple break, gas arc blowing, magnetic blast, and multiple breaks arc blowing. The arc extinguishing principle is to lengthen the arc to increase the arc voltage. As the switching current increases, it becomes more and more difficult to extinguish arc simply by elongating arc and new types of switches need to be considered.

The volume and weight of vacuum circuit breakers are 30% less than that of air switches with the same breaking capacity. The contact resistance of vacuum circuit breaker is $\mu\Omega$ level, and power consumption is less than SSPC. Compared with the external environment, the vacuum arc chamber has negative pressure and good safety. The diffused vacuum arc has low energy, fast dissipation rate when the current is over zero, and the recovery speed of the post-arc dielectric strength is kV/s [6], making the vacuum switch very suitable for the working condition of high frequency. The core of vacuum switch is vacuum arcing chamber, and the design and improvement of vacuum switch revolve around the optimization of arcing performance. The analysis of the complex physical process inside the vacuum arc is not only a hot research in the field of electrical apparatus but also the foundation and key to improve the arc extinguishing ability of the switch.

1.3.2 Research Status of Vacuum Circuit Breaker

When a pair of contact separate in vacuum, current will shrink to a point or several points on the contact surface at the instant of the separation, showing the resistance increased and temperature rise rapidly until the metal evaporates from the contact. During this period, the electric field intensity is extremely high, forming strong electron field emission and resulting in clearance breakdown, thus vacuum arc is formed. Unlike other gas arc plasma, which is formed by ionization of gas molecules or atoms, vacuum arc plasma is provided by melting, evaporation, and ionization of electrode materials. The influence of the electrode process makes the research of vacuum arc theory complicated. According to the electrode process, vacuum arc is usually divided into cathode spot area, arc column area, and anode area.

The cathode spots are some bright areas covered on the surface of the cathode contact. On the one hand, the cathode spots emit electrons to maintain the continuous current of the vacuum arc. On the other hand, the cathode spots also provide metal vapor to the contact gap and ionize the particles emitted by the cathode to maintain the continuity of conductive dielectric between electrodes, which are the source of all physical phenomena of vacuum arc.

Arc column region refers to the discharge region between the cathode spot and the anode sheath layer. The main function is to conduct arc current. The vacuum arc is formed by the fission of the liquid metal bridge during the contact separation and develops outwards from the initial position. After that, the arc column will show different forms according to the current value.

According to the discharge degree of anode area, from weak to strong, the modes of vacuum arc can be divided into a diffuse arc (low current), footpoint and plume (intermediate current), plus anode spot, and intense arc (high current) [7]. In the case of low current, the source of the vacuum arc is the metal vapor provided by the randomly moving spots on the cathode surface. When the metal vapor leaves the emission zone, it is partly ionized once or twice, and partly ionized by the electrons from the cathode. Under these conditions, the anode acts essentially as an electron collector. It collects enough electrons from plasma to maintain the current. Ideally, if the current of plasma is equal to the desired current of anode surface, no anode voltage drop will appear. However, the following conditions will increase the energy in the anode surface: Some high energy positive ions in the cathode emission zone can also reach the anode, so the electron current must increase correspondingly. The anode also receives neutral particles of metal vapor, and the condensation coefficient is less than 1. The anode also receives radiant energy from the cathode and the plasma. In low current conditions, the anode could transmit the energy to achieve an energy balance. However, when the vacuum arc current is very high, too much energy input to the anode will melt the anode, and there will be spots or blocks on the surface of the contact, which are anode spots. If the current does not decrease in a short time, the anode can still evaporate metal particles at the moment when the current passes zero, then the excess metal vapor will slow down the recovery of the dielectric, resulting in the failure of break. Therefore, the anode spot is one of the important factors restricting the current breaking capacity of vacuum circuit breaker.

1.3.2.1 Magnetic Field Control Arc Technology

As mentioned above, when the short circuit current exceeds a certain value, the vacuum arc will be generated and accompanied by anode spots. Thus the contact surface will be seriously ablated, and a large amount of metal vapor will evaporate, and the recovery strength of the post-arc dielectric will be reduced, which will greatly affect the breaking capacity of vacuum switch. To solve this problem, researchers came up with a way to control the arc using magnetic field. The desired magnetic field is generated in the contact gap by changing the current direction by means of cutting the chute and adjusting the contact point on the electrode. The magnetic field is used to control the arc to reduce the contact ablation and improve the current breaking capacity. Around 1952, TMF method was proposed. The idea is to use the TMF coil to drive vacuum arc for high-speed motion, in order to avoid continuous heating on one point and reduce the ablation of anode, thereby preventing serious melting on the surface of the contact and improving the breaking capacity of vacuum switch. In 1958, Schneider from General Electric Company proposed a spiral groove contact structure, which firstly realizing TMF control. In 1962 Lake and Reece proposed the cup contact structure. In 1972, Hundstad from Westinghouse Electric proposed the folded petal-type contact structure [8].

In the 1970s, AMF method was proposed to improve the breaking capacity of vacuum switches. The TMF allows the arc to move rapidly under the action of the magnetic field, avoiding continuous heating of a part of the contact surface to reduce ablation. The design idea of AMF is exactly the opposite: By stabilizing the arc and reducing diffuse, the arc is evenly distributed on the contact surface to ensure the minimum arc energy and weaken the ablative effect. It is proved by experiment and application that the AMF can improve the threshold current of arc from diffuse state to intense state effectively, which is the significance for the development of vacuum circuit breaker. It is also found in the experiment that even at high current, a strong enough AMF can keep the arc diffused, and the cathode spots of the arc can be evenly distributed on the contact surface without serious local melting. Moreover, the arc voltage is low, and the arc energy is small, and the post-arc dielectric strength recovers quickly [9]. At present, the breaking capacity of vacuum circuit breaker with AMF structure has reached 200 kA/12 kV and still has not reached the limit. The application of AMF technology is gradually developing towards high voltage and ultra-high voltage. Most of the vacuum switches adopt AMF contact structure, except for plate contact and TMF contact in small capacity working conditions.

The application of AMF is mainly achieved by the external winding coil of the external permanent magnet or the AMF structure contact [10]. The early studies were carried out by adding coils outside the vacuum switch, and the winding coils were connected to DC or AC source to generate magnetic field. When DC source is connected to the coil, the change of contact diameter and distance has no influence on the strength and distribution of magnetic field, and the AMF is approximately uniform in the whole contact gap. The AMF in the contact gap is related to the amplitude of the AC current when the external winding coil is connected to the AC source. Due to the eddy current effect of the contact, the alternating AMF will lag, which can be avoided by controlling the establishment time of the external AMF. Sekikawa studied arc burning time and movement law of arc root on contact surface under different magnetic field intensity with external coil [11]. Yoshihiko used the external coil to generate AMF with variable magnetic induction intensity B and changed the switching current I, observing the arc shape, cathode spot, and anode melting under different B/I [12]. Zhai used permanent magnets to generate external applied magnetic field and set different magnetic arc blowing intensity by changing the gap between permanent magnets and contact [13]. As the AMF is generated by external magnetic field, it is easy to adjust parameters and obtain experimental data and phenomena under different magnetic field conditions. When the strength of AMF exceeds a certain constant threshold, the anode spot can be prevented at a certain current, and the threshold will be higher for an alternating AMF than for a constant AMF. Schulman studied the critical AMF in which the vacuum arc turns to the diffusion state under the constant and alternating AMF, and found that the threshold of the alternating AMF is about 1.2 times that of the constant AMF [14].

The practical method of applying AMF technology is to design vacuum switch with AMF contact structure. In these contacts, the magnetic field is generated by the alternating current of the main circuit flowing through the contact coil, and the magnetic field is synchronized with the current of the main circuit. At the same time, the alternating current will generate an eddy current field on the contact surface, and the eddy current effect will make the AMF behind the current. When the current passes zero, the AMF still exists, which is unfavorable to the diffusion of the remaining plasma between the arc gap. Liu and Wang studied this problem and calculated the magnetic field in the cup-shaped vacuum arc chamber by using the 3-dimensional finite element method. The results showed that the hysteresis time of the AMF is different, which is related to the position of the radial direction. The hysteresis time of AMF is closely related to design parameters, such as distance, diameter, blade thickness, and contact material [15–17]. At present, the research of vacuum arc in AMF has been from the initial study of external characteristics to the discussion of microscopic and internal mechanisms.

1.3.2.2 Vacuum Arc in AMF

Kimblin found that the peak arc voltage decreased by about 30% under a 20 mT/kA AMF in vacuum arc chamber with contact diameter of 70 mm [18]. Under the condition of the same structure and size, the current interruption capacity is increased by 30%. This indicated that the constraint of AMF on arc can reduce the input energy and increase the threshold of current generating anode spot. Gundlach believed that when the arc voltage was at the minimum, the vacuum arc was in the form of a completely diffused arc and the current channels formed by a single cathode spot were independent and parallel to each other [19]. Yanabu found that the "optimal AMF" of minimum arc voltage can be obtained with the increase of contact diameter under the condition of constant contact spacing [20]. Wang found that the relationship between arc voltage and AMF was approximately shaped as "L", indicating that when the intensity of AMF exceeded a certain critical value, the arc voltage could be kept at relatively small amplitude and arc voltage would not increase significantly even if the intensity of AMF increased [21]. Due to the constraint of AMF, not only the arc voltage decreased, but also the noise component of arc voltage decreased. Morimiya studied the power frequency vacuum arc, and the results showed that when the current was constant, with the increase of AMF strength, the high-frequency components of arc voltage decreased, and the arc voltage curve become smooth [22]. Taylor found that the amplitude of arc voltage spike can be reduced by increasing the AMF in the pulling arc experiment [23]. Xiu and Liu studied the vacuum arc behavior under the AMF contact and observed that the amplitude and duration of arc voltage noise began to increase when the peak current increased from 11.3 to 17 kA and 22.7 kA [24]. The relation curve between magnetic induction intensity of typical AMF and arc voltage is shown in Fig. 1.8.

According to the previous description, the relationship between voltage and current of vacuum arc presents positive volt-ampere characteristics, and in the process of arc burning, the vacuum arc shows two significantly different forms: diffused state and concentration state. With the increase of arc current, when a certain value (7–10 kA for Cu contacts) is reached, the arc will change from the diffused state to the concentration state, which will make the contact surface melt and lead to the failed interruption [25]. However, in the actual arc process, the arc will take on many other forms, which are related to arc mode, current level, contact distance, and other factors. Heberlein used high-speed photography to photograph the vacuum arc shape during the contact breaking process in the detachable vacuum arc chamber [26]. The experimental results show that the arc current, the contact spacing, and the instantaneous current value at the moment of contact separation will influence the evolution of arc shape during the process of breaking. Heberlein also observed that there was

Fig. 1.8 Relationship between arc voltage and AMF (distance is 30 mm)

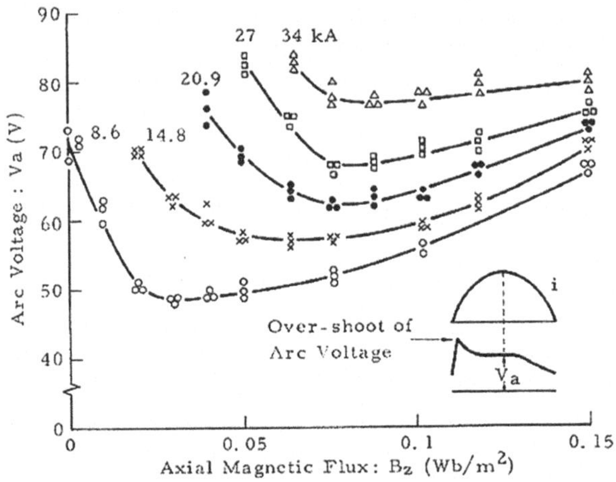

a distinct abrupt shift from the intense evaporation anode injection to the formation of anode spots in high current vacuum arc in the AMF. The experimental comparison showed that strong anode activity occurred at the boundary and the moved to the place with long-distance and large current in AMF. The AMF can maintain the diffused arc at a higher current intensity.

Schulman also believed that the evolution of arc in the process of contact opening was affected by the instantaneous value of opening distance and current [27]. When the instantaneous current value was less than 7 kA at the moment of separation, a transient bridge column arc was formed after the initial molten metal bridge, and its radius kept increasing until a diffused arc was formed. The time interval increased with the increase of the instantaneous current value, but it was always less than 1 ms. When the instantaneous value of the current at the moment of contact separation was between 7 and 15 kA, the radius of the bridge column arc remained stable when it increased to a certain extent, forming a diffused columnar arc. The radius increased with the increase of the instantaneous value of the current and the contact distance at the moment of contact separation, but it was always smaller than the contact radius. In the arc process, when the instantaneous value of the current exceeded 16 kA, the diffused state arc suddenly changed into the intense columnar arc. As the instantaneous value of the current dropped below 10 kA, the intense columnar arc evolved into the diffused state arc. When the instantaneous current value at the moment of contact separation was greater than 15 kA, the vacuum arc evolved directly from the bridge columnar arc to the intense columnar arc. In the case of relatively small opening distance, the formation of anode spot may lead to the formation of jet cylindrical arc, whose shape was like two cones intersecting at the top. After the jet arc was formed, when the instantaneous current exceeded 36 kA, the arc radius would increase. If the distance was further increased, then anode jet and cathode jet would be separated, forming anode jet arc. Schulman made a further study on this experimental phenomenon, and the analysis showed that the transition from bridge cylindrical arc to other arc forms was related to the pressure inside the arc at the moment of transition. When pressure $P > 1.1 \times 10^5$ Pa, the arc was directly transformed to diffused columnar arc.

Voshall studied the breaking capacity of the vacuum plate contact for power frequency AC current. The relationship between the breaking capacity and the contact material, current change rate, rise rate of recovery voltage (RRRV) and contact structural parameters were derived [28].

$$I_{\mathrm{r.m.s}} = \frac{A_{\mathrm{T}}ec}{8\pi\omega\tau\beta KLd^2}\exp[Kc(L+R)/2R(\mathrm{RRRV})L^{1-\gamma}] \tag{1.1}$$

where $I_{\mathrm{r.m.s}}$ was the effective value of the ultimate interrupting current, and A_{T} aws the surface area of the cylinder determined by the contact radius R and the contact opening distance L, and β was the atomic energy loss when each electron reaches the anode, and K was the constant, and d was the atomic diameter of the metal vapor, and RRRV was the rise rate of recovery voltage. From Eq. (1.1), it can be seen that: the faster the RRRV was, or the higher the frequency was, the smaller the current interrupting capacity was. The above conclusions were based on the following assumptions: no contact melting on the contact surface, no influence of anode spots, uniform and isotropic density, the same temperature of all neutral particles between contacts, constant distance during voltage recovery.

Murano used contact with AMF structure to perform an actual short circuit breaking experiment in a detachable vacuum arcing chamber [29]. The following results and conclusions were obtained: when the opening distance of the contact was 10 mm and 40 kA current was interrupted, the chamber could withstand a recovery voltage of 60 kV. For vacuum circuit breakers with AMF, there is no limit on the interrupted current within the normal rated interrupting current range. The range of voltage levels could be improved by selecting the appropriate contact material. It was considered that the AMF type vacuum circuit breaker has the possibility to be applied to high-voltage levels.

Yanabu made a series of experimental studies on the AMF contact in the dismountable vacuum chamber and the actual AMF contact, respectively, and believed that an appropriate magnetic field could limit the arc plasma and minimize the arc energy during arc burning [9, 20]. The uniform distribution of arc spots on the contact surface was found, which proved the arc energy input evenly. From the point of interruption, the uniform distribution of energy is undoubtedly beneficial. This is because the reduction of low energy of arc can avoid the melting of anode, so that the density of plasma will not be too high locally when the current passed zero, eliminating the possibility of local electron emission. The relationship between the interrupted current and contact diameter was summarized as follows:

$$I_{\mathrm{r.m.s}} = \frac{kT}{\pi E\omega d^2 s}\left(1 + C_{\mathrm{i}}\frac{R}{L}\right)^2 \tag{1.2}$$

where, $I_{r.m.s}$ was the effective value of the maximum interrupted current, and E was corrosion rate; d was the diameter of metal atoms, and C_i was the mediation coefficient of contact surface, and S was a constant between 1 and 2, and ω was angular frequency, and K was Boltzmann constant, and T was the density of metal atoms, and R was the contact diameter, and L was the distance. When R was far less than L, the influence of R on current was weakened. The experimental results showed that the $I_{r.m.s}$ was proportional to the 1.4 power of R. It was compared with the breaking capacity of spiral groove contact (the $I_{r.m.s}$ is proportional to the 1.1 power of R, and the current limit is 40 kA). In addition, Yanabu also conducted a series of experiments on the vacuum chamber with the same structure. The switching current started from 60 kA and increased 20 kA each time until to 200 kA. There was only one interruption at the second half-wave. Accordingly, Yanabu believes that 200 kA was still not the breaking limit of the AMF chamber.

Liu compared the vacuum arc characteristics of the cup-shaped AMF contacts with diameters of 48 mm, 58 mm, and 66 mm, with a distance of 8 mm and 14 mm respectively [30]. Through experiments, the relationship between the diameter and the distance ratio (*D*/*g*) and the limit breaking current was discussed, and the limit breaking current under the AMF contact satisfied the following relation:

$$I = k(D/g) + b \tag{1.3}$$

where k and B were constant. Equation (1.3) showed that the limit interrupted current was proportional to the contact diameter at a certain opening distance. When the contact diameter was fixed, the limit interrupted current was inversely proportional to the distance. The experimental results and arc images showed that the arc shape become more diffuse with the contact diameter increasing. When the contact diameter was fixed, the shape of arc in small diameter contacts was diffused as the distance increases, and the arc voltage remained unchanged. However, for the arc in large-diameter contacts, the shape was concentrated with the increase of the distance, and the arc voltage increased with the increase of the distance.

Slade and Smith compared interruption performance, advantages, and disadvantages of AMF and TMF contacts in short distance and long arc time [31]. The AMF contact had good ablative ability and good interruption performance (the voltage level can reach 72 kV) under high current. The TMF could drive the motion of the arc column, which made the TMF contact perform better and have the better ablative ability in the case of long arc time. However, the voltage level could reach 27 kV. Moreover, the TMF contact also had disadvantages including low ablative ability in the case of high current, which was not suitable for the high voltage. Through experimental comparison, the AMF reduced the input energy by stabilizing the arc, so that the AMF contact can perform better in the short arc time.

Based on the research and development experience, Renz introduced the influence of structural parameters on interruption capacity and compared the advantages and disadvantages of AMF and TMF contacts in terms of interruption capacity, contact resistance, size, and production cost in different voltage levels [32]. The criterion for measuring the interruption capacity was the ratio of the maximum charge of the arc to the melting area of the effective contact surface. The comparison results show that compared with the AMF contact, the distance had more serious influence on the breaking capacity for TMF contact. AMF contact was more suitable for high-voltage and high current systems.

In the above reference, researchers have proved that the AMF is beneficial to improve the current breaking capacity through experiment and theoretical analysis. However, it was found that the AMF still existed in the contact gap after the arc was extinguished, which was unfavorable to the arc extinction. This was because the AMF would obstruct the diffusion of the remaining plasma and slow down the rising rate of the recovery strength of the dielectric. When the AC current passed through the contact, an alternating AMF would be generated. However, due to the eddy current effect, the AMF would lag behind the current to a certain phase, which might lead to the existence of a magnetic field after the current zero, affecting the current interruption. For this reason, the contact was grooved to reduce the eddy current. One form of AMF contact, the "multipole AMF", is a big improvement in reducing eddy currents. Kurosawa made a comprehensive analysis of the characteristics of this contact [33]. It was found that the magnetic field intensity was high at the peak current, and the magnetic field intensity was very weak when the current passed zero, which greatly improved the eddy current effect. In addition, he also studied the process and other technical problems in this type of AMF chamber. It was found that the current interruption capacity was related to the arc energy density on the surface of the contact, and was proportional to the square of the contact diameter. Liu conducted two-dimensional simulation of current zero hysteresis in AMF [15, 16]. The simulation results showed that structural parameters such as contact distance, blade thickness, contact material, coil thickness, and height would affect the hysteresis time of magnetic field. Shi studied the power frequency vacuum arc in irregular AMF [34]. They believed that the ideal AMF distribution at current zero should satisfy the requirement, strong at contact edge, and low in the middle. The magnetic field distribution and current breaking capacity of the cup-shaped contact with iron core were studied. It was found that the irregular AMF distribution improved the current breaking capacity effectively.

1.3.2.3 High-Frequency Vacuum Arc

At present, the research on vacuum arc is mainly in power frequency, while the research in high-frequency vacuum (several hundred kHz) arc is mainly on the characteristics of high frequency arc generated in the process of post-arc reignition. Due to the increase in frequency, compared with the power frequency arc, the arc time arc is shortened, and the current change rate $\mathrm{d}i/\mathrm{d}t$ at current zero and the rise rate of recovery voltage $\mathrm{d}u/\mathrm{d}t$ increase, and the eddy current effect is more obvious, which will affect the arc breaking.

Glinkowski studies the characteristics of high-frequency short arc, proving the negative impact by $\mathrm{d}i/\mathrm{d}t$. At the same time, he found the hysteresis phenomenon of voltage when the current crossed zero, that is, the arc voltage lasted for a period of time after the arc current crossed zero. In his opinion, the pause time was caused by the ion current in the arc gap affected by di/dt, which was more likely to cause arc reignition. Moreover, it was proposed that the transient recovery voltage had an adverse effect on arc breaking and the di/dt before current zero and the initial du/dt were important factors affecting the high-frequency short arc breaking [35].

Lindmayer analyzed the influence of circuit parameters and contact materials on high-frequency vacuum arc reignition. The oscillation circuit was used to generate a vacuum current of 100–600 kHz, and the reignition voltage after current zero at different frequencies was analyzed. He concluded that there were two mechanisms of high-frequency arc reignition, cold arc clearance reignition, and reignition under the influence of plasma. The trend of arc reignition increased with the increase of di/dt and current frequency. The voltage of the reignited arc was related to the contact material and increased in the following order, Cu, Cr, and CuCr25 alloy [36]. Heyn conducted further experimental research on the reignition characteristics of the contact material and tested the reignition voltage of CuCr25 alloy contacts in the case of power–frequency 50 Hz and high frequency respectively. It was also found that the reignition voltage of CuCr25 alloy contacts was the highest [37].

Smeets mainly studied the phenomenon of multiple reignitions of high-frequency vacuum arcs at 200 kHz, and pointed out that transient recovery voltage was an important factor determining the breaking capacity in the case of small contact distance. The effects of distance (0.1–1 mm), circuit parameters, and contact materials (Cu, CuCr, and AgWC) on high-frequency arc breaking were studied [38].

Zalucki studied the vacuum arc with a frequency of 5.9–60 kHz. He used a cup contact with a diameter of 80 mm, with an opening distance of 0.3 mm. The amplitude of the first half-wave current was 1–14.5 kA, and the recovery voltage rise rate was 16 kV/μs. The results showed that the arc reignition voltage depended on the amplitude and frequency of the current, the initial current value, and the discharge mode before the half-wave. The reignited voltage had three basic distribution types which are determined by the current amplitude at a given frequency. In the range of 10–60 kHz, the limit breaking current was inversely proportional to the square of the current frequency. In addition, the voltage/current characteristics of the vacuum arc at 50 Hz, 150 Hz, 900 Hz, and 6 kHz were studied. It was found that the noise of arc voltage gradually appeared and the amplitude increased with the increase of current at a given frequency and distance. The noise amplitude of arc voltage increased when the frequency increased [39].

Hardt studied the dynamic arc voltage/current characteristics of 50 Hz and 7 kHz in TMF and evaluated the energy loss of high-frequency vacuum arcs. The contact was of TMF structure, with 45 mm diameter and contact material CuCr25, and distance was 1, 3, and 6 mm. The experimental results showed that the dynamic characteristics of the cold arc clearance at high frequency were similar to those at power frequency [40].

1.3.2.4 VF Vacuum Arc

The arc evolution, arc motion, and arc characteristics of VF (IF, 360–800 Hz) vacuum arc are just at the beginning.

Wang was the first to research aviation VF vacuum arc and he studied the characteristics of VF vacuum arc in AMF [41]. The commercial cup-shaped AMF contact was used in the experiment. The diameter of the contact was 66 mm. The experiment was carried out at a small distance (2–4 mm). The effective values of current ranged from 1.6 to 19.6 kA, and the frequency ranged from 400 to 800 Hz. By studying the limit value of the interrupted current in the cup-shaped AMF interrupter at different frequencies, the inverse relation between the current and the reciprocal of frequency was obtained. The arc energy characteristics in the arc process were deeply discussed, and the relationship between arc energy and anode ablation was explored. The arc forms were divided into three types: free expansion state, transition state, and diffusion state. When analyzing the arc images with constant current and increasing frequency, it was found that the transition state arc appeared more concentrated appearance with increasing frequency. The arc voltages at turning points in the three arc states were studied, and it was considered that both the arc voltages and the average half-wave input power increased linearly with the increase of current value and frequency. The importance of transition arc on anode ablation was confirmed by the study of anode phenomenon. After analyzing the energy characteristics of the VF arc, the characteristic of the transition arc at different distances was summarized. The reasons for the ablation on contact surface were analyzed: firstly, the intensed arc caused by the anode and cathode at small distances; secondly, the increased input power caused by the increased frequency.

Then, Zhu studied the characteristics of VF vacuum arc in TMF [42]. The interruption experiment of small diameter TMF vacuum arc chamber was carried out at the frequency of 400–800 Hz. The magnetic field measurement platform and eddy current field model of cup-shaped TMF contact was established, and the magnetic field distribution of vacuum arc chamber under different frequencies was measured and simulated. It was concluded that with the increase of current frequency, the magnetic induction intensity of TMF decreased, and the hysteresis time of TMF increased. The simulation was carried out at different frequencies when the structure parameters and materials of the TMF contact changed. The dynamic volt-ampere characteristics of the arc at different frequencies were compared, and the arc development process was divided into three stages: the initial arc burning stage, the transition stage, and the current dropping stage. When the frequency increased, the area covered by the volt-ampere characteristic curves was larger. By observing and analyzing the cause of voltage noise, it was found that arc motion was not the cause of arc voltage noise, but when the cathode spot occupied the whole cathode surface, the arc voltage produced high-frequency noise. It is also found that the evolution of arc shape at different frequencies existed in three modes: expansion arc column, moving arc column, and diffuse arc. The research results of arc motion in TMF contact showed that the average velocity of the arc column increased with the increase of frequency under the same interrupted current. It was found that with the increase of frequency, the number of split arcs decreased obviously. The paralleled split arc could reduce the arc voltage. As the collapse time of arc plasma was inversely proportional to the current rise rate, the collapse time of arc was shorter and the arc was more likely to concentrate at the same current than at the power frequency. In the extended arc phase, when the frequency was higher, the arc diameter became smaller, and the average half-wave input power would be concentrated in a smaller area.

When the current frequency is increased, the half-wave time is shortened and the characteristics of VF vacuum arc will change. Researchers established a continuous transition model considering the factors of arc process. It was found in the range of 1000–2500 Hz, growth of sheath was slow, which was not conducive to arc breaking [43]. In the experiment, the author of this book found that for VF vacuum arcs with short distances, no obvious concentration state was observed. Most arcs were in the diffusion state, and no serious large-area melting occurred on the contact surface [44]. We have designed asynchronized AMF system. By adjusting the excitation intensity, the influence of AMF on the characteristics of the IF vacuum arc could be analyzed [45]. And we have already observed that the arc diffuse uniformly in the AMF contact, while the arc moves vigorously in the plate contact. According to Bennet's pinching model, the pressure gradient was the main reason that affected the shape of VF vacuum arc and the formation of plasma jet [46]. The unique characteristics of the VF vacuum arc were as follows: with the increase of frequency, on the one hand, the current change rate di/dt at zero increased, making it difficult for the arc to extinguish. On the other hand, the contact anode spot was not significant during arc burning, and the source of metal vapor would change after arc.

With the increase of frequency, the interruption process of VF vacuum arc will be much more difficult. In Ref. [47], researchers believed that when the opening distance is fixed, the reduction of arc time and current frequency could improve the recovery strength of dielectric. Studies in Ref. [48] showed that when the current frequency is 500 Hz, with the increase of current, AMF could accelerate the diffusion of the cathode spots, and recover the strength of dielectric, and shorten the breaking time. The author of this book found that with the increase of frequency, the breaking capacity of VF vacuum switch decreased. Combined with the ablation degree of contact surface and electric field concentration factors after the failed interruption, the reason why the heavy ignition point of VF vacuum arc appeared at the edge of the contact was given [49]. The existence of metal particles was more unfavorable to the recovery of dielectric strength. When the VF vacuum arc reignited, it was found that metal droplets were splashed outward from the gap [50]. The splashing process was quantitatively analyzed, and it was found that the droplet injection velocity was 10–20 m/s, and the internal pressure gradient of the arc gap was about 7.9×104 Pa/mm. Moreover, after the current passed zero, the metal droplet could disappear for a longer time than the half-wave time, which was more unfavorable for the breaking than 50 Hz. Therefore, it was considered that the particles of metal droplet could affect the breakdown process as the metal vapor source [51]. However, the motion model of metal particles was only in the plane of arc image, and cannot reflect the spatial distribution of the pressure in the arc chamber. The splashing behavior of metal particles had time–space characteristics and was closely related to the time, position, and probability of post-arc breakdown under the action of recovery voltage. If we want to deeply understand the physical nature of the breakdown of VF vacuum arc, it is a key problem to analyze the influence of metal particles on the dielectric recovery process from the perspective of microscopic mechanism.

1.4 The Contents of This Book

This chapter is the Introduction. In this chapter, we mainly introduce the background and significance of this book, and introduce the research status of aviation power supply systems, aviation electrical apparatus, vacuum arc theory, and breaking technology.

Chapter 2 is an experiment and Simulation Platform of Variable Frequency power system. In this chapter, we describe the establishment of the experimental and simulation platform of variable intermediate frequency power system.

Chapter 3 is Characteristics of Contacts of Variable Frequency Vacuum Circuit Breaker. The influence of the contact materials in AMF on the characteristics of VF vacuum arc voltage and arc shape is described through comparative experiments in this chapter.

Chapter 4 is Frequency Characteristics of Variable Frequency Vacuum Arc. In this chapter, we mainly discuss the arc characteristics of AMF and plate contacts at different frequencies and current levels.

Chapter 5 is Post-arc Breakdown in Variable Frequency Vacuum Arc. In this chapter, we mainly describe the phenomenon and mechanism of VF vacuum arc breakdown.

References

1. Zhuoran Zhang, Li Yu, Jincai Li, Jiawei Lu. Key technologies of advanced aircraft electrical machine systems for aviation electrification [J]. Journal of Nanjing University of Aeronautics and Astronautics, 2017, 49(5): 622–634.
2. Long Cheng, Fanghua Zhang, Shuo Liu, Zehua Zhang. Configuration method of hybrid energy storage system for high power density in more electric aircraft [J]. Journal of Power Sources, 2020, 445: 227322.
3. Yuanjun Zhou, Yong Wang, Huifen Dong. Electric power systems of civil aircraft [M]. Shanghai: Shanghai Jiao Tong University Press, 2015: 1–11.
4. B. Sarlioglu, C. T. Morris. More electric aircraft: review, challenges, and opportunities for commercial transport aircraft [J]. IEEE Transactions on Transportation Electrification, 2015, 1(1): 54–64.
5. Zhihong Xu. Theory fundamentals of electrical apparatus [M]. Beijing: Machinery Industry Press, 2014: 168–171.
6. Zhenxing Wang, YingsanGeng, Zhiyuan Liu. Stepwise behavior of free recovery processes after diffused vacuum arc extinction [J]. IEEE Transactions on Dielectrics and Electrical Insulation, 2012, 19(2): 582–590.

7. H. C. Miller. Anode modes in vacuum arcs: update [J]. IEEE Transactions on Plasma Science, 2017, 45(8): 2366–2374.
8. H. Schellekens. 50 years of TMF contacts design considerations [C]. 23rd International Symposium on Discharge and Electrical Insulation in Vacuum, Bucharest, Romania, 2008, 1: 95–98.
9. S. Yanabu, E. Kaneko, H. Okumura, et al. Novel electrode structure of vacuum interrupter and its practical application [J]. IEEE Transactions on Power Apparatus and Systems, 1981, PAS-100(4): 1966–1974.
10. J. A. Foosnaes, W. G. J. Rondeel. Vacuum arc subjected to an axial magnetic field [J]. Journal of Physics D: Applied Physics, 1979, 12(11): 1867–1871.
11. SekikawaJunya, Kubono Takayoshi. Motion of break arcs driven by external magnetic field in a DC42V resistive circuit [J]. IEICE Transactions on Electronics, 2008, E91-C(8): 1255–1259.
12. Yoshihiko Matsui, Akira Sano, Hideki Komatsu, et al. Vacuum arc phenomena under various axial magnetic field and anode melting [C]. 24th International Symposium on Discharge and Electrical Insulation in Vacuum, Braunschweig, Germany, 2010: 324–327.
13. GuofuZhai, Xue Zhou, Wenying Yang. Experiment on DC inductive arcs driven by axial and transverse magnetic fields [J]. Transactions of China Electrotechnical Society, 2011, 26(1): 68–74.
14. M. B. Schulman, P. G. Slade, J. V. R. Heberlein. Effect of an axial magnetic field upon the development of the vacuum arc between opening electric contacts [J]. IEEE Transactions on Components, Hybrids and Manufacturing Technology, 1993, 16(2): 180–188.
15. Zhiyuan Liu, Zhongyi Wang, Jimei Wang, ShixinXiu. Study on phase shift time of cup-type axial magnetic field contact [J]. High Voltage Apparatus, 2004, 40(2): 87–90.
16. Zhiyuan Liu, KesongXie, Zhongyi Wang, Jimei Wang. Analysis of 3D eddy current field in cup-type axial magnetic field vacuum interrupter [J]. Advanced Technology of Electrical Engineering and Energy, 2004, 23(2): 26–37.
17. Zhongyi Wang, Zhiyuan Liu, Xuan Zhang, Jimei Wang. Comparison of axial magnetic field characteristics of 5 axial magnetic field vacuum interrupter contacts [J]. Advanced Technology of Electrical Engineering and Energy, 2006, 25(1): 21–25.
18. C. W. Kimblin, R. E. Voshall. Interruption ability of vacuum interrupters subjected to axial magnetic fields [J]. Proceedings of the Institution of Electrical Engineers, 1972, 119(12): 1754–1758.
19. H. C. W. Gundlach. Interaction between a vacuum arc and an axial magnetic field [C]. 8th International Symposium on Discharges and Electrical Insulation in Vacuum, Albuquerque, NM, 1978: A2.1–A2.11.
20. S. Yanabu, S. Souma, T. Tamagawa, et al. Vacuum arc under an axial magnetic field and its interrupting ability [J]. Proceedings of the Institution of Electrical Engineers, 1979, 126(4): 313–320.
21. Yi Wang, Jimei Wang. Investigation of vacuum arc subjected to axial magnetic field [J]. Proceedings of the CSEE, 1987, 7(5): 34–41.
22. O. Morimiya, S. Sohma, T. Sugawara, et al. High current vacuum arcs stabilized by axial magnetic fields [J]. IEEE Transactions on Power Apparatus and Systems, 1973, PAS-92(5): 1723–1732.
23. E. D. Taylor, P. G. Slade, M. B. Schulman. Transition to the diffuse mode for high-current drawn arcs in vacuum with an axial magnetic field [J]. IEEE Transaction on Plasma Science, 2003, 31(5): 909–917.
24. ShixinXiu, Zhiyuan Liu, Jimei Wang. Vacuum-arc behaviors of a coil-type axial-magnetic-field contact at contact gap of 60 mm [J]. IEEE Transaction on Plasma Science, 2008, 36(1): 208–214.
25. M. P. Reece. The Vacuum switch [J]. Proceedings of Institute of Electrical Engineers, 1963, 110(4): 793–811.
26. J. V. R. Heberlein, J. G. Gorman. High current metal vapor arc column between separating electrodes [J]. IEEE Transactions on Plasma Science, 1980, PS-8(4): 283–288.

27. M. B. Schulman, P. G. Slade. Sequential modes of drawn vacuum arcs between butt contacts for currents in the 1 kA to 16 kA range [J]. IEEE Transactions on Components, Packaging, and Manufacturing Technology Part A, 1995, 18(2): 417–422.
28. R. E. Voshall. Current interruption ability of vacuum switches [J]. IEEE Transactions on Power Apparatus and Systems, 1972, PAS-91(3): 1219–1224.
29. M. Murano, S. Yanaba, T. Tamagawa. Interruption ability of vacuum arc [C]. 3rd International Conference on Gas Discharges, 1974, 118:648–652.
30. Zhiyuan Liu, Shaoyong Cheng, Yuesheng Zheng, et al. comparison of vacuum arc behaviors between axial-magnetic-field contacts [J]. IEEE Transactions on Plasma Science, 2008, 36(1): 200–207.
31. P. G. Slade, R. K. Smith. A comparison of the short circuit interruption performance using transverse magnetic field contacts and axial magnetic field contacts in low frequency circuits with long arcing times [C]. 21st International Symposium on Discharges and Electrical Insulation in Vacuum, Yalta, Crimea, 2004, 2:337–340.
32. Renz. The application of contacts optimization for the Longitudinal and transverse magnetic fields in vacuum arc extinguish chambers [J]. Northeast Electric Power Technology, 2002, 23(1): 23–25.
33. Y. Kurosawa, H. Sugawara, Y. Kawakubo, et al. Vacuum circuit breaker electrode generating multi-pole axial magnetic field and its interruption ability [J]. IEEE Transactions on Power Apparatus and Systems, 1980, PAS-99(6): 2079–2085.
34. Zongqian Shi, ShenliJia, Jun Fu, et al. Axial magnetic field contacts with nonuniform distributed axial magnetic fields [J]. IEEE Transactions on Plasma Science, 2003, 31(2): 289–294.
35. M. Glinkowski, A. Greenwood. Some interruption criteria for short high-frequency vacuum arcs [J]. IEEE Transactions on Plasma Science, 1989, 17(5): 741–743.
36. M. Lindmayer, E. D. Wilkening. Influence of circuit parameters and contact materials on the reignition of high-frequency vacuum arcs [J]. IEEE Transactions on Components, Hybrids, and Manufacturing Technology, 1990, 13(1): 69–73.
37. D. Heyn, M. Lindmayer, E. D. Wilkening. Effect of contact material on the extinction of vacuum arcs under line frequency and high frequency conditions [J]. IEEE Transactions on Components, Hybrids, and Manufacturing Technology, 1991, 14(1): 65–70.
38. R. P. P. Smeets, T. Funahashi, E. Kaneko, et al. Types of reignition following high-frequency current zero in vacuum interrupters with two types of contact materials [J]. IEEE Transactions on Plasma Science, 1993, 21(5): 478–483.
39. Z. Zalucki. Interrupting capacity of vacuum interrupters depending on the frequency of current [J]. IEEE Transactions on Plasma Science, 1993, 21(5): 494–500.
40. N. Hardt, M. Heimbach, H. Bohme, et al. The dynamic voltage/current characteristics of vacuum arcs after breakdown at currents in the lower kHZ-range [J]. European Transactions on Electrical Power, 2002, 12(5): 321–326.
41. Jing Wang. Properties of intermediate-frequency vacuum arc on axial-magnetic-field contacts [D]. Beijing: Beihang University, 2011.
42. Liying Zhu. Study on intermediate-frequency vacuum arc behaviors of transverse-magnetic-field contacts [D]. Beijing: Beihang University, 2013.
43. Can Ding, Zhao Yuan, Junjia He. Effect of vacuum arc cathode spot distribution on breaking capacity of the arc-extinguishing chamber [J]. Japanese Journal of Applied Physics, 2017, 56(10): 106001.
44. Yuan Jiang, Jianwen Wu. Effects of contact materials and diameters on characteristics of intermediate-frequency vacuum arc in axial magnetic field [J]. Proceedings of the CSEE, 2015, 35(20): 5367–5374.
45. Yuan Jiang, Jianwen Wu, Wei Tang. Excitation system of external axial magnetic field in intermediate-frequency current interruption experiment [J]. Transactions of China Electrotechnical Society, 2015, 30(9): 39–45.
46. Yuan Jiang, Jianwen Wu, Suliang Ma, Mingxuan Chen, Bowen Jia. Appearance of vacuum arcs in axial magnetic field and butt contacts at intermediate frequencies [J]. IEEE Transactions on Plasma Science, 2019, 47(2): 1405–1412.

47. Taotao Qin, Enyuan Dong, Guixin Liu, Jiyan Zou. Recovery of dielectric strength after DC interruption in vacuum [J]. IEEE Transactions on Dielectrics and Electrical Insulation, 2016, 23(1): 29–34.
48. Zhanqing Chen, XiongyingDuan, Minfu Liao, Jiyan Zou, Ruitong Liu. Influences of arc parameters on the repeated interruption performances of laser triggered vacuum switch [J]. Transactions of China Electrotechnical Society, 2019, 34(21): 4501–4507+4600.
49. Yuan Jiang, Qing Li, Jiarui Cui, Jianwen Wu, Bowen Jia. Re-ignition of intermediate frequency vacuum arc at axial magnetic field [J/OL]. Transactions of China Electrotechnical Society, https://kns.cnki.net/kcms/detail/11.2188.TM.20191211.1117.002.html.
50. Yuan Jiang, Jianwen Wu. Interruption phenomenon in intermediate-frequency vacuum arc [J]. Plasma Science and Technology, 2016, 18(3): 311–318.
51. Yuan Jiang, Yukun Liu, Qing Li, Jiarui Cui, Weicun Zhang, Jianwen Wu, Bowen Jia. Interruption and droplets emission in intermediate-frequency vacuum arc of aviation power supply system [J]. Proceedings of the CSEE, 2020, 40(2): 684–692.

Chapter 2
Experiment and Simulation Platform of Variable Frequency Power System

2.1 Experiment Platform

The main function of the circuit breaker is to remove the fault in the power system automatically. It is an important research content to analyze the mechanism of arc generating in the process of interrupting and improve the current breaking capacity, so the interruption test is the most important experiment [1]. The synthetic circuit is used in the experiment of power frequency. In synthetic experiment, the breaking current and recovery voltage is provided by two power sources. The low-voltage and high current source are for arc burning. Before current zero, the high-voltage power supply is put into at an appropriate time as the transient recovery voltage, making the tested circuit breaker under the high-voltage condition [2]. In order to ensure the equivalence of the synthesis test, the error time of voltage source input is required to be less than 50 μs, which makes it difficult to control.

A similar test platform is needed to generate short circuit current in VF vacuum circuit breaker experiments. After analyzing the characteristics, advantages, and disadvantages of the synthetic circuit and the voltage level of the aviation power supply, a VF vacuum arc generation and observation platform with arc initiation branch is designed in this chapter, as shown in Fig. 2.1. The measure and control system can record the current, voltage, and image of VF vacuum arc.

2.1.1 Power Circuit

The power circuit of VF vacuum arc experiment is composed of LC oscillation circuit and arc initiation branch, using a bidirectional thyristor to replace the ignition ball gap in the synthetic experiment. The platform is simple, easy, and can provide short circuit current with a frequency of 360–850 Hz and peak value of more than 20 kA.

The oscillation circuit is composed of capacitor C_1, inductor L_1, bidirectional thyristor VT_1, vacuum circuit breaker VI, and shunt R_1. L_1 and C_1 are used to

Y. Jiang and Q. Li, *Vacuum Circuit Breaker for Aviation Variable Frequency Power System*, Power Systems, https://doi.org/10.1007/978-981-33-4781-6_2

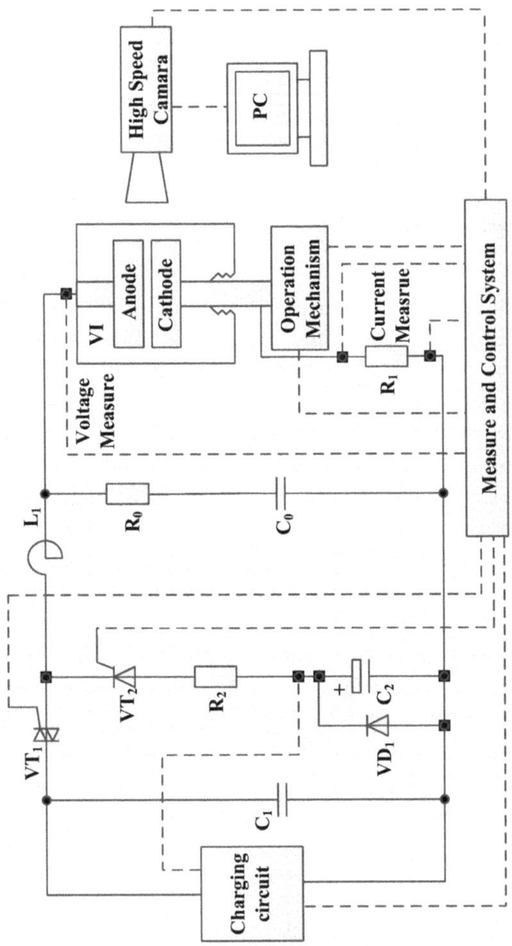

Fig. 2.1 Test platform for VF vacuum circuit breaker

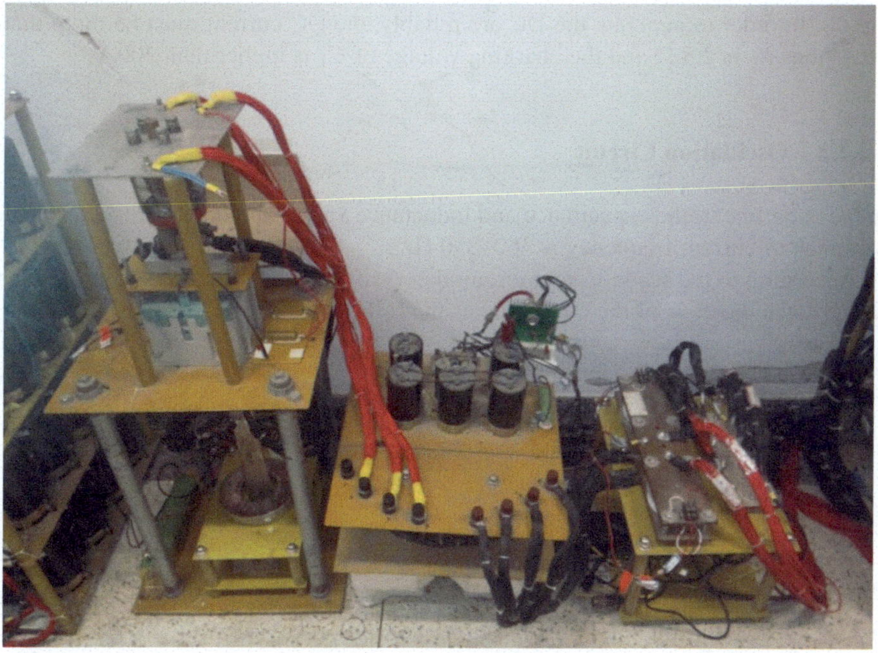

Fig. 2.2 A physical picture of the experimental system

generate the VF current required for the experiment, and VT_1 is the trigger switch for the circuit. The arc initiation branch consists of capacitor C_2, current limiting resistor R_2, thyristor VT_2, and power diode VD_1. It is used to draw and maintain a DC vacuum arc in VI until VF current is introduced. The frequency modulation branch consists of R_0 and C_0, which is applied to adjust the RRRV after current zero. The sequence of experimental operations is as follows. When capacitor C_1 and C_2 are ready to be charged, the VT_2 is first triggered when VI is closed. If the VI contacts open, the DC arc will be generated by the arc initiation branch. Then VT_1 is triggered to introduce VF current as the contacts of VI are open. VT_2 is switched off due to the negative voltage, for the voltage of C_1 is higher than that for C_2. The trigger signal of VT_1 will keep to the end of the experiment. Figure 2.2 shows the physical diagram of the experimental system. From left to right are vacuum arc circuit breaker VI, arc initiating branch, inductor, and bidirectional thyristor VT_1.

The design principles of parameters in each part are described as follows.

2.1.1.1 Arc Initiation Branch

The function of the arcing initiation branch is to draw and maintain the DC arc before VF current is injected. The electrolytic capacitor C_2 discharges through current limiting resistor R_2 when VT_2 is triggered, and the power diode VD_1 is used to protect the C_2. In order to generate the DC arc reliably, the DC current must be more than 80 A, thus R_2 is 2.5 Ω and the charging voltage of C_1 is higher than 200 V.

2.1.1.2 Oscillation Circuit

In the experiment, the capacitance and inductance values should be adjusted so that the range of current frequency is 360–850 Hz.

According to the second-order circuit theory, only if the resistance R, inductance L, and capacitance C of the circuit meets the following relation:

$$R < 2\sqrt{L/C} \tag{2.1}$$

The circuit will oscillate, and the oscillating current can be described as:

$$i(t) = \frac{U_0}{\omega L} e^{-\delta t} \sin(\omega t) \tag{2.2}$$

Among which, U_0 is the voltage of capacitor C, and δ is the time constant, $\delta = R/2L$, and ω is the angular frequency of the second-order circuit, $\omega = \sqrt{1/LC - \delta^2}$.

From the above equations, it can be seen that the resistance R should be as small as possible if the inductance L and capacitance C are determined. Firstly, the oscillation conditions of the circuit should be satisfied. Secondly, the current amplitude should be attenuated slowly but not too fast. However, in practice, the reduction of resistance requires a thick wire, which will increase the cost and installation difficulty for the entire system.

The nonpolar CBB60 capacitor group is selected as the capacitor for the VF oscillating circuit. The epoxy glass cloth material is used to ensure insulation and fire protection and facilitates wiring. The physical object is shown in Fig. 2.3.

The short circuit current of the VF experiment should reach more than 20 kA instantly, and the iron core inductor will be rapidly saturated. The inductance required for the circuit is at the order of μH, depending on the frequency and the capacitance. The hollow inductance is chosen for the experiment in this book for it is not saturated at high current. The following equation is used to calculate the inductance value:

$$L = \frac{0.08 d_{cp}^2 \times n^2}{3d_{cp} + 9H + 10d} \tag{2.3}$$

Among which, d_{cp} is the average winding diameter (unit cm), and N is winding turns, and H is the winding height (unit cm), and d is winding thickness (unit cm). The physical object is shown in Fig. 2.4.

Bidirectional thyristor VT_1 consists of two thyristors in reverse parallel. Due to the short arc time of VF current, the surge working state of bidirectional thyristor (10

Fig. 2.3 Capacitance an oscillating circuit

Fig. 2.4 Inductance in oscillating circuit

times overload) can meet the current requirements in the experiment. Bidirectional thyristor can improve the time control precision significantly.

Chamber is the core device of vacuum circuit breaker, which is responsible for arc interrupting. The gas pressure inside the chamber should be less than 10^{-3} Pa. When the vacuum arc chamber with current is opened by an external mechanism, the current will continue to conduct for a period of time between the upper and lower contacts, and then vacuum arc will appear. The higher the current, the harder it is to interrupt the arc. The shunt resistor is made of constantinite wire with low-temperature coefficient in parallel.

2.1.1.3 Frequency Adjustment Branch

Whether AC arc can be extinguished after current zero depends on the diffusion process of post-arc plasma, which is affected by many factors. The recovery strength of dielectric is an important parameter to describe this phenomenon. In the breaking process, if the recovery voltage is higher than the recovery strength of dielectric, the arc will reignite. In the power frequency switching experiment, the recovery voltage includes two parts: the transient recovery voltage and the power frequency recovery voltage. The transient recovery voltage plays a decisive role in arc extinguishing.

When the power frequency synthesis circuit is used to test the breaking performance of the high-voltage switch, the transient recovery voltage has strict regulations, and the methods to determine the parameters are two-parameter method and four-parameter method. These parameters can be adjusted by changing the resistance and capacitance paralleled with the contact.

At present, there is no unified method for the design and parameters of transient recovery voltage in the VF aviation power supply system. According to the regulation on transient recovery voltage characteristics in the standard of industrial low-voltage electrical apparatus, with the condition of individual motor load (inductive load), the oscillation frequency of transient recovery voltage in this book is calculated according to the following equation:

$$f = 2000 I_{\mathrm{c}}^{0.2} U_{\mathrm{e}}^{-0.8}\,(1 \pm 10\%) \tag{2.4}$$

Among which, f is the oscillation frequency of recovery voltage (unit kHz), and I_{c} is the effective value of current (unit A), and U_{e} is effective value of rated operating voltage (unit V).

2.1.2 *Control System*

Figure 2.5 is the schematic diagram of control system for VF vacuum circuit breaker experiment. PLC and single-chip microcomputers are used to control the action sequence of arc initiation thyristor, operating mechanism, a high-speed camera, and

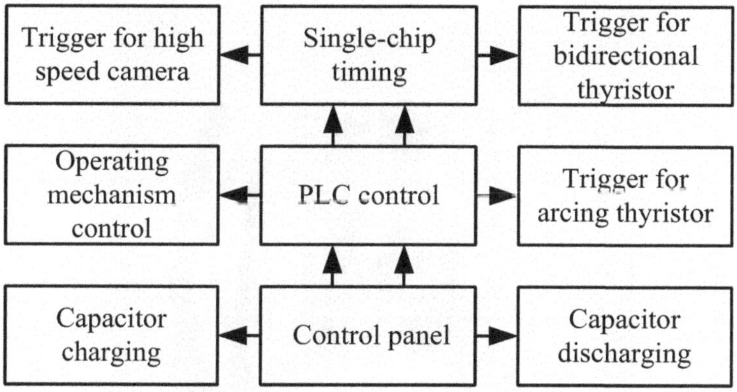

Fig. 2.5 Control system for VF vacuum circuit breaker experiment

bidirectional thyristor in the power circuit. Before the experiment, oscillation capacitor C_1 and arc initiation branch capacitor C_2 are charged by the charging circuit respectively. Sequence of the experiment is as follows. First, arc initiation thyristor VT_2 is triggered, generating DC arc. Then VI is driven by operating mechanism, and 10 ms later the camera system will work. Then VT_1 is triggered to inject VF current while the contacts are fully open. At the end of the experiment, the residual power of the capacitor is consumed by the discharge circuit.

In order to control the action of VI conveniently, contactor is used as the operating mechanism, as shown in Fig. 2.6. The DC control method of AC contactor is used to reduce the error of operation time. After the capacitor is stored, the instantaneous discharge circuit is used to drive the electromagnetic mechanism, and the contact of the VI is opened. Then, the low-voltage DC power supply is used to keep the electromagnetic mechanism in stable suction, which can guarantee the timing control accuracy to the maximum extent.

2.2 AMF Excitation System

For a vacuum circuit with AMF structure, it is necessary to find the optimal AMF value to prevent anode spots. If the contact structure is designed, the parameter value of *B*/*I* is determined. With the same current, the magnetic field intensity will be the same, so it is hard to analyze the effect of magnetic field on vacuum arc. When an external magnetic field is generated by external constant DC source or permanent magnet, the changes of contact diameter and opening distance would not affect the intensity and distribution of AMF which is approximately evenly distributed in the contact gap. A scientific method of establishing external magnetic field is to generate a time-varying AMF, whose amplitude can be adjusted, and the phase and frequency

Fig. 2.6 The contactor electromagnetic operating mechanism

are synchronized with the main circuit current. In this section, an external magnetic field system is designed, which can produce an alternative VF AMF [3].

2.2.1 *Structure*

The external excitation AMF system is shown in Fig. 2.7. The oscillating circuit, including the capacitor C, inductance L, bidirectional triode thyristor VT, vacuum interrupter VI, and resistor R_d, is applied to generate VF current and vacuum arc. In the dotted box is the external AMF excitation system. Helmholtz coil consisted of winding coils W_1 and W_2 with equal turns, is to establish AMF, which is set outside the VI. Current transformers (CTs) provide excitation current synchronous with the VF current in the power circuit for Helmholtz coil. Due to the magnetic induction intensity B can be adjusted by the numbers of CTs, the effect of the excitation system can be measured by the ratio of B/I. The excitation system should meet the following requirements:

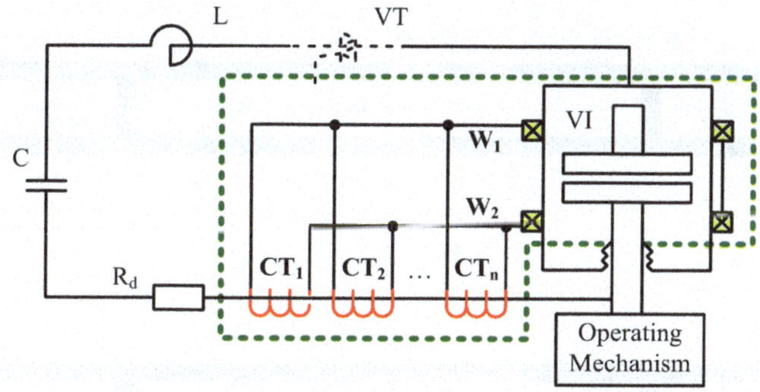

Fig. 2.7 External AMF excitation system for VF vacuum circuit breaker

(1) For each CT, the value of B/I is designed as 4 mT/kA.
(2) According to the national standard for current transformer in China, the phase difference of current between the primary side and the secondary side should be less than 10′ (0.2 grade).
(3) It is necessary to ensure the core of CT unsaturation when the VF current reaches the maximum peak value (20 kA).

2.2.2 *Modeling and Design*

The main components of the excitation system are the Helmholtz coil and CT.

2.2.2.1 Helmholtz Coil

For the convenience of description, the parameters of Helmholtz coil are explained as follows.

Helmholtz coil is made up of two coils in series, as shown in Fig. 2.8. The number of turns in a coil is N, and the radius of wire is r_1. The average radius of the coil is R, which is as same as the center distance between the two coils. The equivalent inductance of Helmholtz coil is L_H. The excitation current is I_0, and the related primary current in the power circuit is I. Among which, N and R are the parameters to be decided.

According to Biot–Savart's law, the magnetic induction intensity $B(x_P)$ at any point P in the coil axis is:

$$B(x_P) = \frac{\mu_0 N I_0 R^2}{2(R^2 + x_P^2)^{3/2}} + \frac{\mu_0 N I_0 R^2}{2(R^2 + (R - x_P)^2)^{3/2}} \tag{2.5}$$

When P is in the middle of the axis, that is $x_P = R/2$, then we have

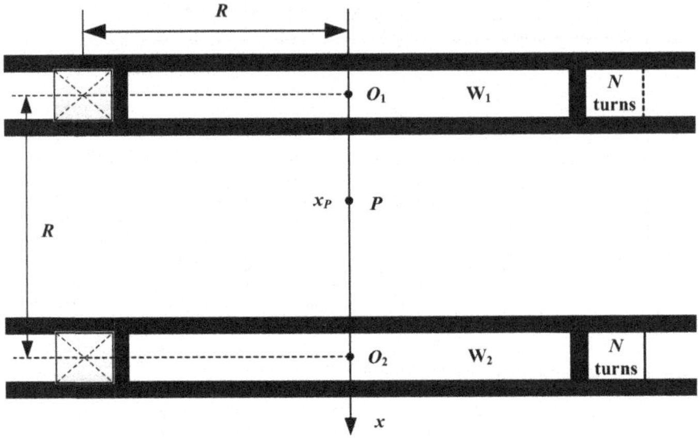

Fig. 2.8 Structure diagram of the Helmholtz coil

$$B\left(\frac{R}{2}\right)=\left(\frac{4}{5}\right)^{3/2}\frac{\mu_0 N I_0}{R}=\frac{K\mu_0 N I_0}{R} \tag{2.6}$$

In Eq. (2.6), the number of turns of Helmholtz coil can be expressed as:

$$N=\frac{BR}{K\mu_0 I_0}=\frac{Rk}{K\mu_0}\cdot B/I \tag{2.7}$$

Among which, $k = I/I_0$ is the current conversion ratio of CT, and the average radius R is determined according to the volume of VI. The equivalent inductance of Helmholtz can be expressed approximately as,

$$L_h=\frac{\Psi}{I_0}\approx\frac{2NB\pi R^2}{I_0}=2\pi K\mu_0 RN^2 \tag{2.8}$$

2.2.2.2 CT

For the primary side of the CT, the current is I, and the voltage is U, and the number of turns is 1. For the secondary side of the CT, the current is I_0, and the number of turns is N_0, and the radius of wire is r_2, and the current density is J. The mean radius of the core is R_0, and the sectional area of the core is S, and the magnetic induction intensity is B_{Fe}.

We have Eq. (2.9) for J and N_0:

$$\frac{N_0}{1}=k=\frac{I}{I_0}=\frac{I}{4\pi r^2 J} \tag{2.9}$$

When designing CT used in power frequency, current density J is usually selected as 5–10 A/mm^2. Due to the short period of VF current, the current time in the experiment is in milliseconds. For example, the period of current of 400 Hz is 2.5 ms. Then the current density J can be increased appropriately. With the experimental verification, in a short period of time (less than 10 ms), the wire with radius of 0.25 mm can withstand a current density of 120 A/mm$_2$.

The electromagnetic coupling relationship between the primary side and secondary side of the CT can be represented by the equivalent circuit in Fig. 2.9a. R_1 and X_1 are the resistance and leakage reactance in the primary side. R_m and X_m are the excitation impedance. R_2 and X_2 are the resistance and leakage reactance in the secondary side calculated to the primary side. R_h and X_h are the impedance of the load coil calculated to the primary side. I is the current in power circuit, and I_m is the current in excitation system, and I_h is the current in secondary side.

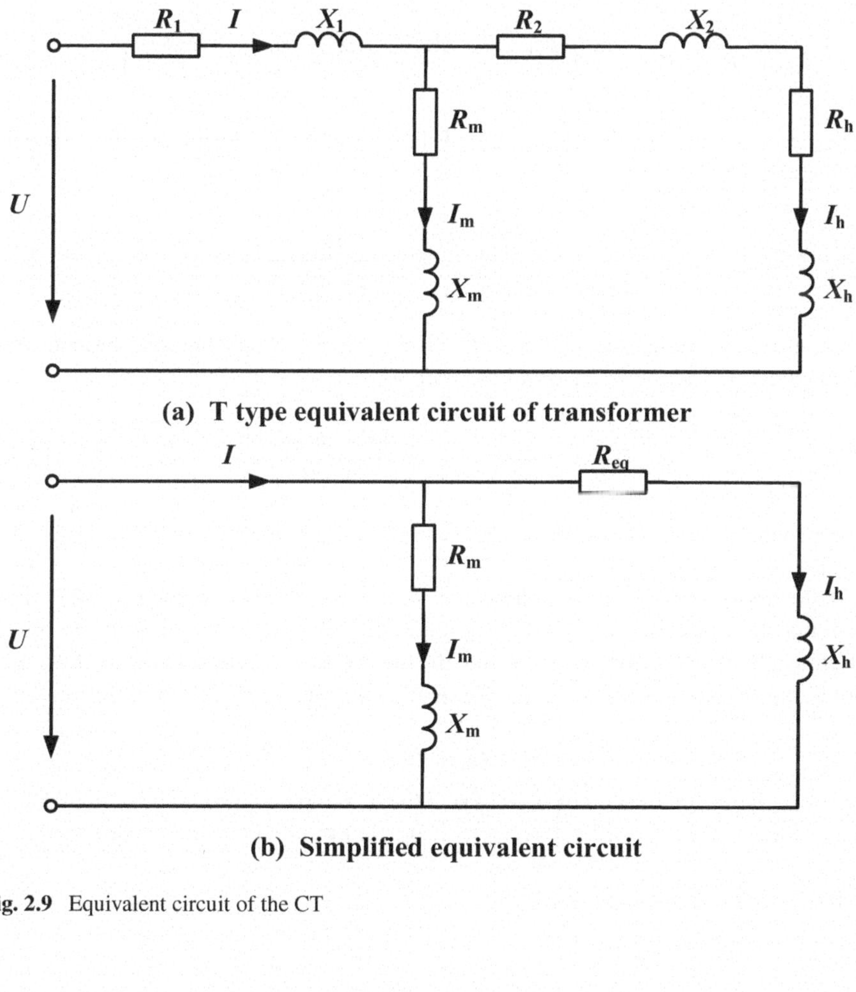

Fig. 2.9 Equivalent circuit of the CT

And for convenience of analysis, we make the following simplification. The primary resistance R_1 and the leakage inductance X_1 and X_2 are neglected. The resistance R_2 and R_h are combined into equivalent resistance R_{eq}. The simplified equivalent circuit is shown in Fig. 2.9b, and the calculation relationship is as follows.

The excitation reactance X_m is:

$$X_{\mathrm{m}} = \omega L_{\mathrm{m}} = \frac{\omega \mu_{\mathrm{r}} \mu_0 S}{2\pi R_0} \tag{2.10}$$

The inductance X_h of the load coil is:

$$X_{\mathrm{h}} = \frac{\omega L_{\mathrm{h}}}{k^2} = \frac{2\pi K \mu_0 \omega R N^2}{k^2} \tag{2.11}$$

The total equivalent resistance in the secondary side is:

$$R_{\mathrm{eq}} = R_2 + R_{\mathrm{h}} = \frac{1}{k^2}\left(\frac{\rho_{\mathrm{Cu}} \cdot 2N \cdot 2\pi R}{2\pi r_1^2} + \frac{\rho_{\mathrm{Cu}} \cdot N_0 \cdot (2a + 2b)}{2\pi r_2^2}\right) \tag{2.12}$$

Since the relative permeability, μ_r, of the core in CT is very large, for example, $\mu_r = 4000$. We have:

$$\frac{X_{\mathrm{m}}}{X_{\mathrm{h}}} = \frac{\frac{\omega \mu_{\mathrm{r}} \mu_0 S}{2\pi R_0}}{\frac{2\pi K \mu_0 \omega R N^2}{k^2}} = \frac{S k^2}{4\pi^2 K R R_0 N^2} \mu_{\mathrm{r}} \tag{2.13}$$

It can be concluded that most of the current flow through the load branch. And the magnetic induction intensity of the core is:

$$B_{\mathrm{Fe}} = \frac{U}{4.44 \cdot f \cdot S} = \frac{4K\pi^2 \mu_0 R I N^2}{4.44 k^2 S} \tag{2.14}$$

When the current is at the peak, the magnetic induction intensity of the core is the greatest.

In the above analysis, only inductances of the Helmholtz coil and CT are considered and the resistance of the wire is ignored. However, if we analyze the phase difference between the primary side and the secondary side, the resistance of the wire should not be ignored. We have the following current phasor equation in Fig. 2.9:

$$\begin{cases} \boldsymbol{I} = \boldsymbol{I}_{\mathrm{m}} + \boldsymbol{I}_{\mathrm{h}} \\ \boldsymbol{U} = \boldsymbol{I}_{\mathrm{m}} \cdot (R_{\mathrm{m}} + jX_{\mathrm{m}}) \\ \boldsymbol{U} = \boldsymbol{I}_{\mathrm{h}} \cdot \left(R_{\mathrm{eq}} + jX_{\mathrm{h}}\right) \end{cases} \tag{2.15}$$

By further arrangement, we have:

$$\frac{\boldsymbol{I}_{\mathrm{h}}}{\boldsymbol{I}} = \frac{R_{\mathrm{m}} + jX_{\mathrm{m}}}{R_{\mathrm{m}} + R_{\mathrm{eq}} + j(X_{\mathrm{m}} + X_{\mathrm{h}})} \tag{2.16}$$

The phase difference between the primary side and the secondary side can be expressed as,

$$\Delta\theta = \arctan\frac{X_{\mathrm{m}}\cdot R_{\mathrm{eq}} - R_{\mathrm{m}}\cdot X_{\mathrm{h}}}{X_{\mathrm{m}}(X_{\mathrm{m}} + X_{\mathrm{h}}) + R_{\mathrm{m}}\cdot(R_{\mathrm{eq}} + R_{\mathrm{m}})} < \arctan\frac{R_{\mathrm{eq}}}{X_{\mathrm{m}} + X_{\mathrm{h}}} = 2.16' \tag{2.17}$$

This result is less than $10'$, which is the design requirement of VF CT.

The pictures of Helmholtz coil and CT are shown in Fig. 2.10. In order to avoid eddy current in core skeleton, non-conductive and non-magnetic epoxy materials are used as the skeleton of Helmholtz coil.

2.2.3 Results and Discussion

2.2.3.1 Test for Phase Difference

In order to verify the phase difference between the primary side and the secondary side of the CT, the oscillation circuit shown in Fig. 2.7 is applied. The current frequency can be adjusted from 360 to 800 Hz by C and L. The measured data are shown in Fig. 2.11. All of the results are less than $10'$.

2.2.3.2 Test for Steady Performance of AMF

In order to evaluate the steady performance of external AMF and the magnetic field errors caused by different currents, the Helmholtz coil is tested. The schematic diagram of steady performance testis shown in Fig. 2.12.

The magnetic induction intensity at the center of the Helmholtz coil is measured by a tesla meter, and the result is shown in Table 2.1 and Fig. 2.13.

Among which, the theoretical value of B is obtained according to Eq. (2.5).

2.2.3.3 Test for Dynamic Performance of AMF

A schematic diagram of test for dynamic performance of AMF is shown in Fig. 2.14. The magnetic induction intensity of AMF can be adjusted by changing the number of CTs, and we applied 2 CTs in this test. The test coil W_{T} is settled at the center of the Helmholtz coil. For W_{T}, the turn of coil is N_{T}, and the cross-sectional area is S_{T}, and the measured voltage is U_{T}.

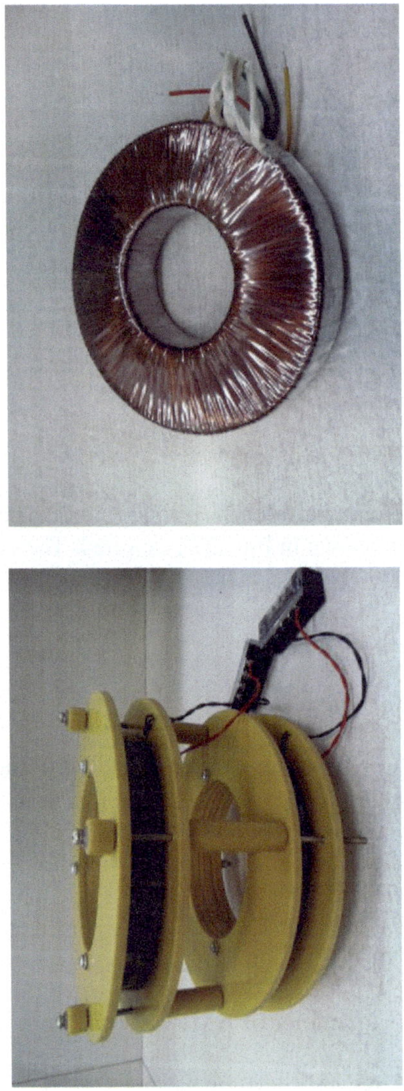

(a) Helmholtz coil

(b) IF current transformer

Fig. 2.10 Picture of Helmholtz coil and CT

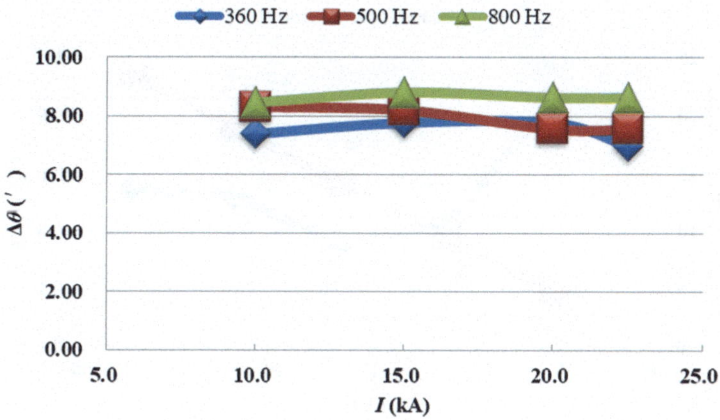

Fig. 2.11 Results of test for phase difference

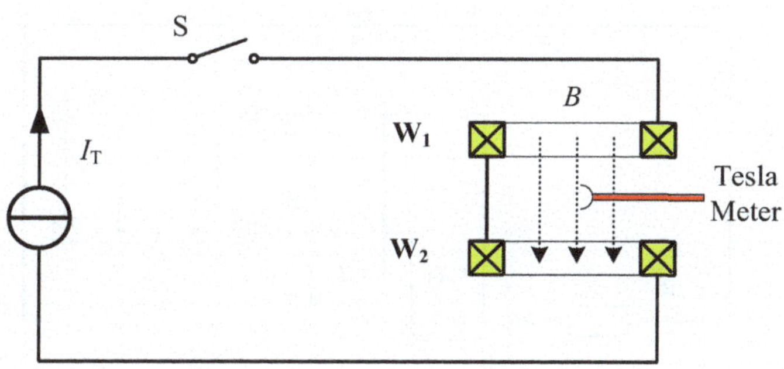

Fig. 2.12 Schematic diagram of the test for steady-state of AMF

Table 2.1 Test result of steady performance

No.	Content			
	I_T (A)	B_{Theory} (mT)	B_{Actual} (mT)	Error (%)
1	5.41	4.67	4.63	0.86
2	11.05	9.54	9.49	0.42
3	16.59	14.32	14.27	0.35
4	22.09	19.07	18.97	0.52
5	27.46	23.70	23.57	0.55

Curves of the dynamic test are shown in Fig. 2.15. Where, U_I is the voltage of the R_d, which is synchronous with the current I in power circuit, and U_T is the measured voltage of W_T.

As can be observed, there is a "steep" rise at the beginning of U_I, indicating that the rise rate of current is very large at the start, leading to drastic changes in the measured voltage U_T. After the initial time, the current is approximately sinusoidal,

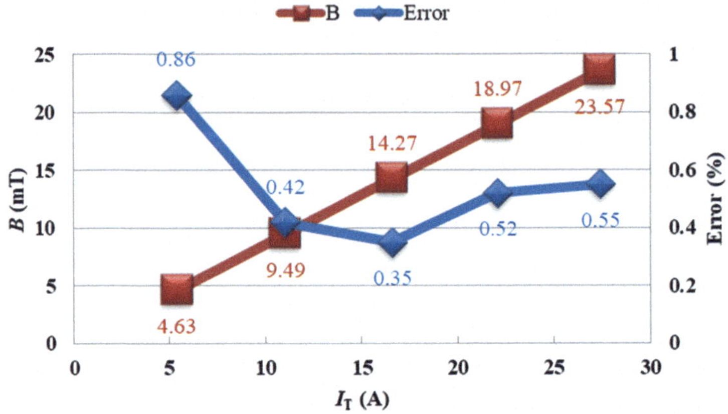

Fig. 2.13 *B* and error of the steady performance of AMF

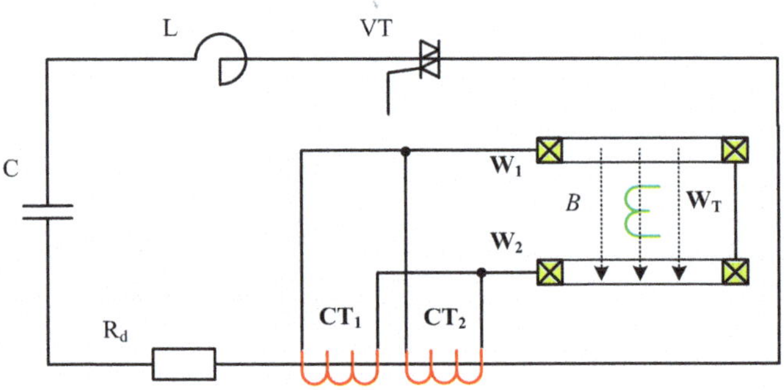

Fig. 2.14 Schematic diagram of the IF dynamic test

Fig. 2.15 Measured curves of the dynamic test

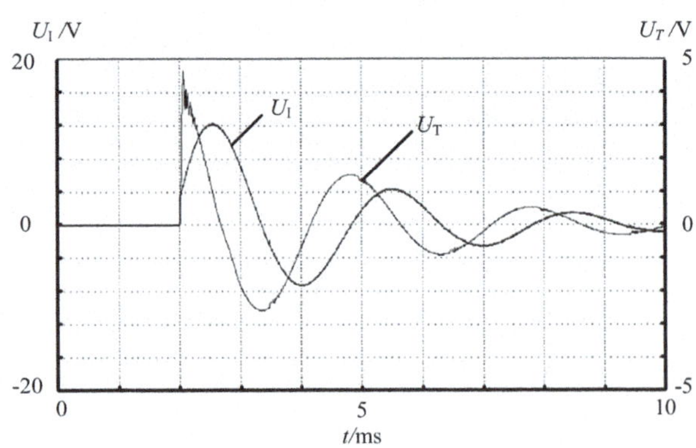

and the voltage tends to be normal. Suppose the current in power circuit is a sine with exponential decay:

$$I = I_{\mathrm{m}} \sin(\omega t + \varphi) \cdot \mathrm{e}^{-t/\tau} \tag{2.18}$$

Then, according to Eq. (2.5), the magnetic induction intensity B at the center of Helmholtz coil is described as:

$$B = \frac{K\mu_0 N I}{R N_0} = B_{\mathrm{m}} \sin(\omega t + \varphi) \cdot \mathrm{e}^{-t/\tau} \tag{2.19}$$

The measured voltage of the test coil W_{T} is:

$$\begin{aligned} U_{\mathrm{T}} &= -\frac{\mathrm{d}(N_{\mathrm{T}} \cdot B \cdot S_{\mathrm{T}})}{\mathrm{d}t} \\ &= -N_{\mathrm{T}} B_{\mathrm{m}} S_{\mathrm{T}} \left[\omega \cos(\omega t + \varphi) \cdot \mathrm{e}^{-t/\tau} - \tau^{-1} \sin(\omega t + \varphi) \cdot \mathrm{e}^{-t/\tau}\right] \\ &= N_{\mathrm{T}} B_{\mathrm{m}} S_{\mathrm{T}} \sqrt{\omega^2 + \tau^{-2}} \sin(\omega t + \alpha) \cdot \mathrm{e}^{-t/\tau} \\ &= U_{\mathrm{m}} \sin(\omega t + \alpha) \cdot \mathrm{e}^{-t/\tau} \end{aligned} \tag{2.20}$$

According to U_{m} and B_{m}, we can obtain:

$$B_{\mathrm{m}} = \frac{U_{\mathrm{m}}}{N_{\mathrm{T}} S_{\mathrm{T}} \sqrt{\omega^2 + \tau^{-2}}} \tag{2.21}$$

And B/I is:

$$B/I = \frac{B_{\mathrm{m}}}{I_{\mathrm{m}}} = \frac{U_{\mathrm{m}}}{I_{\mathrm{m}} N_{\mathrm{T}} S_{\mathrm{T}} \sqrt{\omega^2 + \tau^{-2}}} \tag{2.22}$$

The theory value, the actual value of B/I as well as the errors are displayed in Fig. 2.16.

As can be obtained from the results, the actual value is consistent with the theory value. Even if there is an error caused by the difference between the mathematical model and the actual system. The errors are in an acceptable range.

By theory analysis and experiment, it is shown that the current in power circuit can be induced by the CT. The phase difference between the primary side and the secondary side of the CT is less than 10′. In the center of the Helmholtz coil, AMF is generated and the actual value of B/I is consistent with the theoretical value. The steady-state error of B is less than 1% as well as the dynamic error of B/I is less than 1.5%. As the VF current is up to maximum value (20 kA), the core of CT is not saturated. In addition, the external AMF excitation system can be used for power frequency circuit breaker test, as well as for VF test, which is also the extended content of this work.

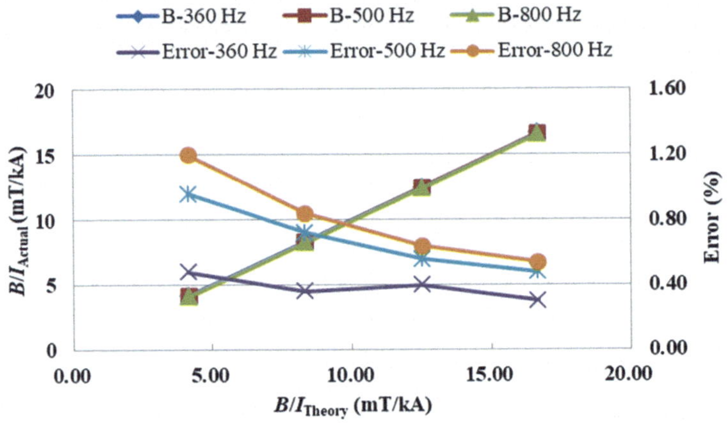

Fig. 2.16 Results of the dynamic test

2.3 Simulation Platform

The vacuum circuit breaker can generate AMF as the current passes through the contact, thus the breaking capacity of vacuum circuit breaker can be improved. As mentioned above, the results of studies have shown that the AMF with appropriate strength can make the arc diffused and evenly distributed on the surface of contact even if in high current mode, reducing the arc energy and the degree of ablation of contact [4, 5].

The structure of cup-type AMF contact is shown in Fig. 2.17. The AMF contact consists of conductive rod, coil, contact blade, and stainless steel support. The arc area is between the two contacts.

For the AMF contact, the eddy current effect will be enhanced as the frequency is increased, with a magnetic field hysteresis which will hinder the plasma diffusion after current zero. As the cycle time of current is changed with the variational frequency, the delay time is not sufficient to explain the hysteresis. Thus, in this chapter, we introduce current phase, θ, to analyze the AMF distribution with the same instantaneous current value while different frequency. For example, θ is $\pi/2$ or 90° when current is positive peak value. The current is given by the following expression:

$$i = \sqrt{2}I_0 \sin\theta = \sqrt{2}I_0 \sin(2\pi f \cdot t) \tag{2.23}$$

where I_0 is the current effective value, and f is the frequency and θ is the current phase.

The AMF B in the contact gap is related to the current magnetic field B_{i} which is generated by the current i, and eddy current magnetic field be which is generated by the eddy current i_{e}. According to Maxwell's equations, when displacement current

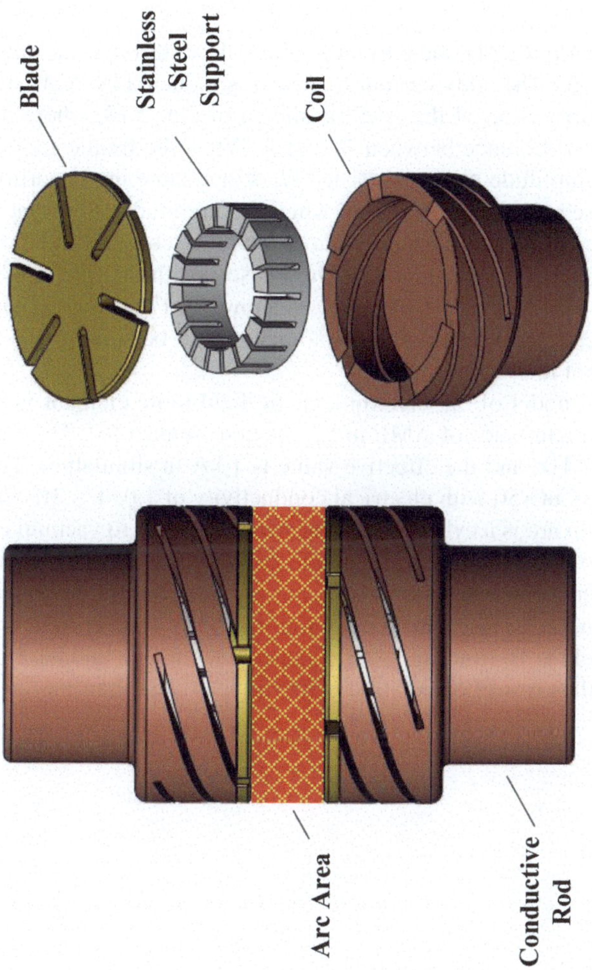

Fig. 2.17 The structure of cup-type AMF vacuum contact

is not taken into account in power circuit, then we have:

$$\begin{cases} \boldsymbol{B} = \boldsymbol{B}_{\mathrm{i}} + \boldsymbol{B}_{\mathrm{e}} \\ \oint\limits_{l} \boldsymbol{B}_{\mathrm{i}} \cdot \mathrm{d}\boldsymbol{l} = \mu_0 i \\ \oint\limits_{l} \boldsymbol{E}_{\mathrm{e}} \cdot \mathrm{d}\boldsymbol{l} = -\int\limits_{s} \frac{\partial \boldsymbol{B}}{\partial t} \cdot \mathrm{d}\boldsymbol{s} \end{cases} \tag{2.24}$$

As known from Eq. (2.24), the current i, which flows through the contacts, is in the same phase as B_{i}. The eddy current i_{e}, which is inducted by AMF, lags behind B by $\pi/2$. The composition of the AMF is shown in Fig. 2.18, where the angle α indicates the phase difference between B_{i} and B. When the frequency of current is constant and the amplitude changes, B_{i} and B_{e} will change in proportion, making the hysteresis phase α constant. However, when the amplitude of current is constant while the frequency increases, the amplitude of B_{i} will be constant and the amplitude of B_{e} will increase. In order to maintain the phasor synthesis, the amplitude of B must decrease as well as α must increase, as shown in Fig. 2.18b. Therefore, the eddy current effect and the hysteresis phenomenon will be strengthened when the frequency of current increases.

In this chapter, model of 3D electromagnetic field finite element is established to analyze the characteristics of AMF in VF vacuum contact [6]. The frequency of current is 360–800 Hz, and the effective value is 1 kA in simulation. The material of contact blade is CuCr50 with electrical conductivity of 1.044×10^7 S/m [7]. We suppose the vacuum arc is a cylinder with the same diameter to vacuum contact and the same height to the separation. The conductivity of VF vacuum arc is 400 S/m according to experimental results: as the diameter of contact is 41 mm and the separation is 3 mm, the average arc voltage is about 50 V when current is 10 kA [8]. The calculation field is 10 times the size of the contact model, and the simulation is terminated when the error of calculation matrix is set to less than 1%.

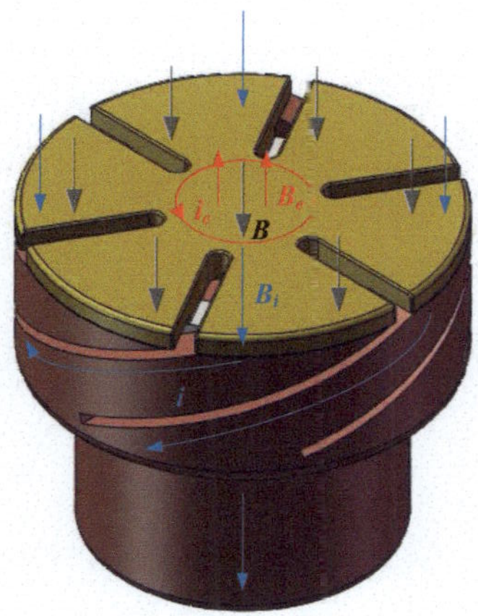

(a) Schematic diagram of the AMF

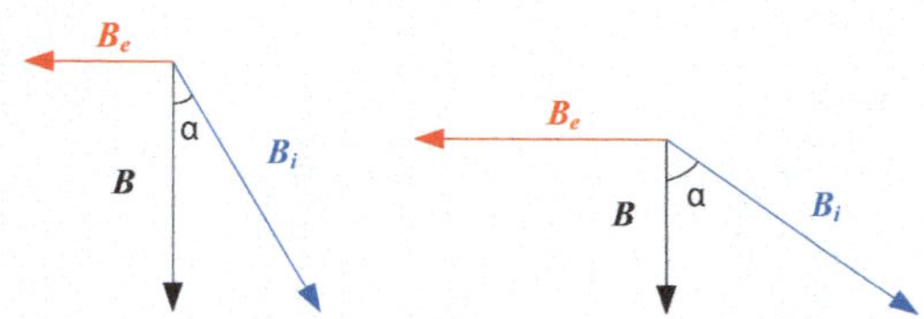

(b1) **(b2)**

(b) Composition of the AMF

Fig. 2.18 The AMF generated by the contacts

References

1. Guozheng Xu, JierongZhag, Jiali Qian, Yulong Huang. Principle and application of high voltage circuit breaker [M]. Beijing: Qinghua Press, 2000.
2. Yuan Jiang, Jianwen Wu, Hefei Wang, FeiGao. Synchronous control system for synthetic test of circuit breaker [J]. High Voltage Apparatus, 2015, 51(8): 152–157.
3. Yuan Jiang, Jianwen Wu, Wei Tang. Excitation system of external axial magnetic field in intermediate-frequency current interruption experiment [J]. Transactions of China Electrotechnical Society, 2015, 30(9): 39–45.
4. Zhiyua Liu, Zhongyi Wang, Xuan Zhang, Jimei Wang. Analysis of axial magnetic field characteristics of coil type axial magnetic field vacuum interrupters [J]. Transactions of China Electrotechnical Society, 2007, 22(1): 47–53.
5. Hui Li, Yuan Jiang, Qing Li, Jianwen Wu, Xiaowu Luo, Bowen Jia, Ziang Tong. Characteristics of axial magnetic field in intermediate frequency for vacuum interrupters: experimental simulation platform design [J]. Vacuum Electronics, 2019(5): 21–27+46.
6. Shenli Jia, Zongqian Shi, Lijun Wang, Zheng Wang, MingzheRong. Experimental investigation of vacuum arcs in nonuniform axial magnetic fields [J]. IEEE Transactions on Plasma Science, 2004, 32(5): 2113–2117.
7. Zhiyuan Liu, Dong Wang, MingzheRong, Jimei Wang, Quan Wang. Comparison of vacuum arc behaviors for slot-type axial magnetic field contacts with and without iron plates [J]. IEEE Transactions on Plasma Science, 2009, 37(8): 1458–1468.
8. Yuan Jiang, Jianwen Wu. Interruption phenomenon in intermediate-frequency vacuum arc [J]. Plasma Science and Technology, 2016, 18(3): 311–318.

Chapter 3
Contacts Characteristics of Variable Frequency Vacuum Arc

3.1 Contacts for the VF Experiment

Diameter and material of contact are important parameters for vacuum circuit breakers. It is of great theoretical and practical value to study arc characteristics under different materials and diameters [1].

In this chapter, experiments with different current have been done in vacuum interrupters with the materials of contact being Cu, Cu-W-WC alloy, and CuCr50 alloy, and with the diameters being 41 mm and 58 mm respectively. Detail of the parameters is shown in Table 3.1.

The effect of materials on arc characteristics and breaking capacity is shown as following: high melting point and steam pressure, good thermal conductivity, and electric conductivity are good for vacuum arc interruption. In addition, low electrical wear rate, anti-fusion welding ability, and strong mechanical resistance are required as well. For vacuum circuit breakers applied in high-voltage circuits, materials also need high breakdown strength. The metallurgical properties of contact materials used in the experiment are shown in Table 3.2 [2, 3].

By comparison, it can be found that there are good thermal and electrical conductivity, low melting point, low hardness, and vapor pressure for Cu. There are low chopping current and the highest vapor pressure at melting point for Cr, which will generate lower noise in arc voltage. And the hardness of metal alloy can be significantly improved by adding Cr. W is a kind of refractory material whose melting point, boiling point, and hardness are very high. There is good corrosion resistance, good wear tolerance and high thermal conductivity for W. WC is with high melting and boiling point, however poor thermal and electrical conductivity.

Figure 3.1 shows the chambers of vacuum circuit breakers used in the experiment. In (a), the material of the contact is CuCr50 alloy, and the diameter of the contact is 58 mm. In (b), the material of the contact is CuCr50 alloy, and the diameter of the contact is 41 mm. In (c), the material of the contact is Cu-W-WC alloy, with a diameter of 41 mm. In (d), the material of the contact is Cu with a diameter of 58 mm.

Y. Jiang and Q. Li, *Vacuum Circuit Breaker for Aviation Variable Frequency Power System*, Power Systems, https://doi.org/10.1007/978-981-33-4781-6_3

Table 3.1 Vacuum circuit breaker for VF vacuum arc experiment

No.	Structure	Diameter of contact (mm)	Material of contact	Main component content (%)
1	AMF	58	CuCr50	Cr: 47–52
2	AMF	41	CuCr50	Cr: 47–52
3	AMF	58	Cu	Cu: 99
4	AMF	41	Cu-W-WC	W: 66–68 Cu: 28–32 C: 1–2

Table 3.2 Metallurgical properties of contact material components

Composition	Melting point (°C)	Boiling point (°C)	Vapor pressure at the melting point (N/m^2)	Thermal conductivity (w/m K)	Electrical conductivity (MS/m)	HV (N/mm^2)
Cu	1083	2570	0.052	394	59.9	1040
Cr	1875	2220	1030	91	10.3	*
W	3410	5927	4.4	167	18	4300
WC	2870	6000	*	29.3	5.2	1620

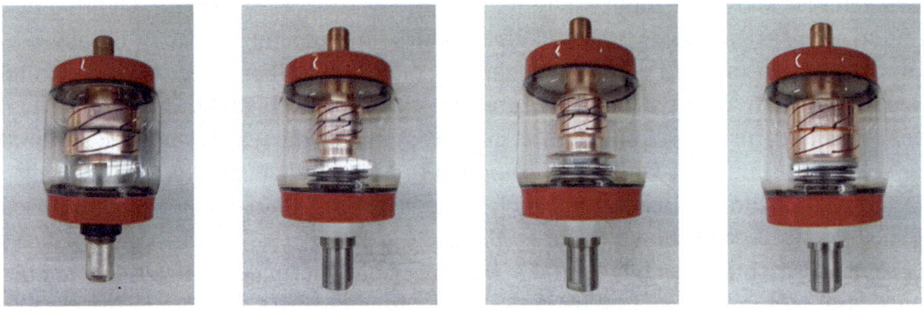

(a) CuCr50, 58 mm (b) CuCr50, 41 mm (c) Cu-W-WC, 41 mm (d) Cu, 58 mm

Fig. 3.1 Vacuum chamber with different contact materials

3.2 Vacuum Arc Influenced by Diameter of Contact

The VF vacuum arc images at current peak time but with different current levels are displayed in Fig. 3.2, with the frequency being 360 Hz and the material being CuCr50. When the peak value of current is 10 and 13.5 kA, the vacuum arc in the two contacts can be interrupted, which is called low current mode. When the peak value reaches up to 20 kA, the arc fails to be interrupted by contact with diameter of 41 mm. And if the peak value reaches up to 22.5 kA, the arc fails to be interrupted

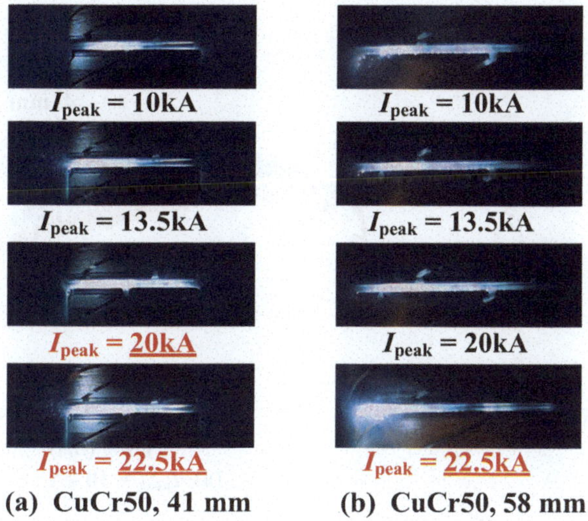

Fig. 3.2 Arc images in CuCr50 contacts with diameters of 41 and 58 mm (360 Hz)

by contact with diameter of 58 mm as well, which is called high-current mode. In Fig. 3.2, it is marked by an underline to distinguish the two modes.

By comparing the arc images in the two contacts, it can be observed that as follows. When in the same current, the vacuum arc between the contacts with a diameter of 41 mm is brighter and more concentrated. Cathode spots can be observed between contacts with a diameter of 58 mm obviously, and the distribution of arc is uniform and the diffusion state is better. According to the brightness of arc images, the input power for contacts of 41 mm is higher. With the increase of current, the arc diameter is increasing and the arc is becoming brighter as well. In the low current mode, the vacuum arc in both of the two contacts is not filling the distance gap fully. While in high current mode, for the contacts of 41 mm, if the current peak is 20 kA (the effective value of current in the first half cycle is 14.3 kA) and 22.5 kA (the effective value of current in the first half cycle is 16 kA), the vacuum arc is full of distance gap. For the contacts of 58 mm, when the current peak is 20 kA, the diameter of the arc is increasing significantly. And if the current peak is 22.5 kA, the vacuum arc is spilled and concentrated, with the brightness enhanced significantly and almost no cathode spots observed obviously. Based on the results of experiment, factors such as the brightness, diameter, and shape of the vacuum arc, are related to whether the vacuum arc is interrupted successfully or not.

The arc voltage at low current mode at 360 Hz is shown in Fig. 3.3. When the peak current value is 10 kA, voltages in two contacts are flat and there is no obvious fluctuation. The maximum arc voltage in 41 mm contact is 36.2 V, higher than the maximum arc voltage of 31.9 V in 58 mm contact. For with same the current, higher voltage also reflects that the arc power in 41 mm contacts is higher. When the peak current is 13.5 kA, arc voltage in the 58 mm contact is still flat with the maximum value 36.5 V. However, there is a voltage spike in the 41 mm contact with the maximum value 56 V. The arc voltage at high-current mode at 360 Hz is shown

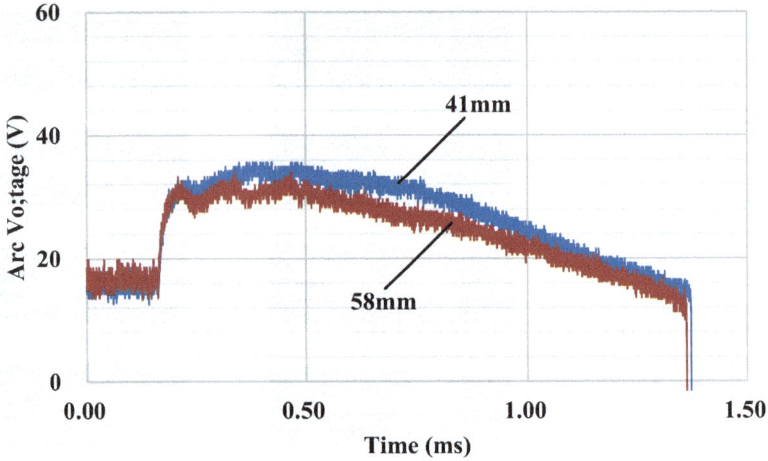

(a) $I_{peak} = 10$ kA

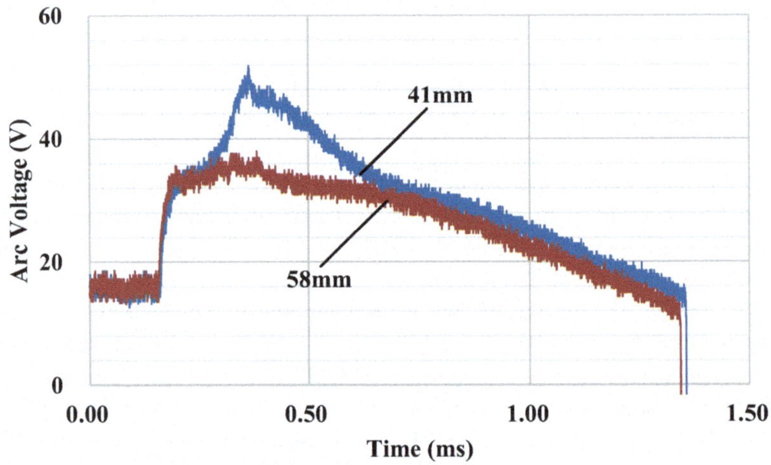

(b) $I_{peak} = 13.5$ kA

Fig. 3.3 Arc voltages of CuCr50 contacts at low current mode

in Fig. 3.4. While the current is 20 kA, there is a voltage spike in the 58 mm contact with the maximum value of 50 V. And the vacuum arc is failed to be interrupted in the 41 mm contact. When the vacuum arc is reignited, voltage is fluctuant. When with the same current, anode is hotter with smaller diameter so it is easier to be melted, thus the energy density must be higher. As a result, reignition is easier to happen. And if the current increases to 22.5 kA, neither of the two contacts can interrupt the vacuum arc.

As mentioned above, when current is the same, the vacuum arc is stable and diffused in larger contacts which performs better at breaking capacity. Firstly, the reason is that the vacuum arc can be stabilized and distributed uniformly by the

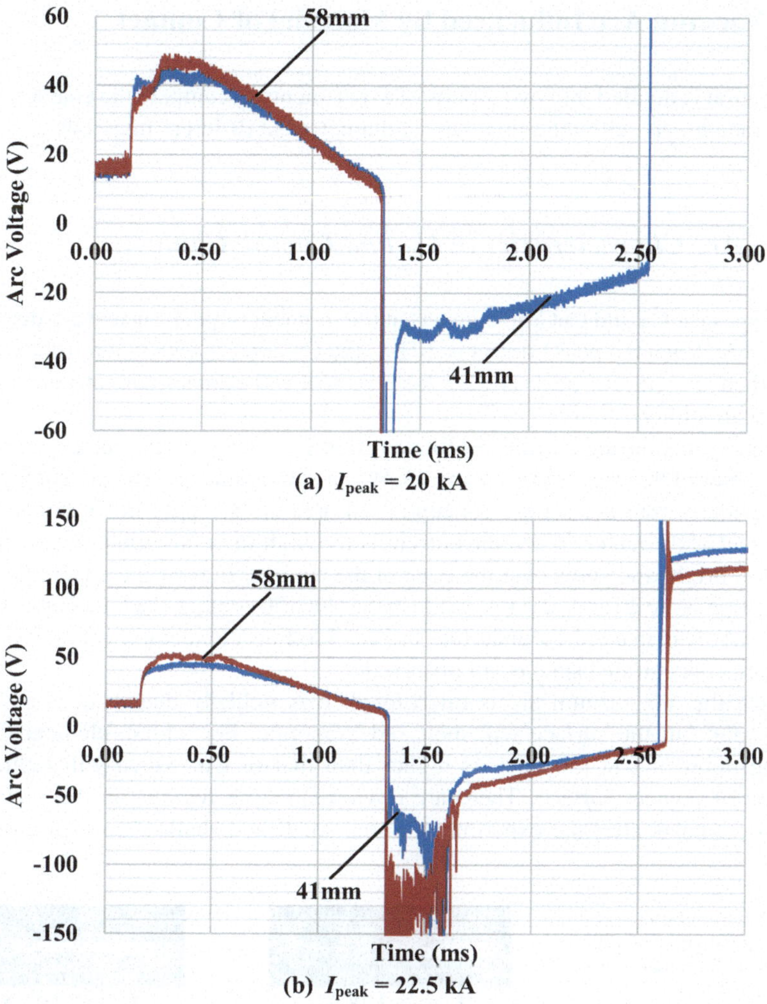

Fig. 3.4 Arc voltages of CuCr50 contacts at high current mode

AMF if the contact is larger, therefore the injected power is minimized and erosion of contact is reduced and diffused vacuum arc is good for the interruption. Secondly, with a larger area, current density in surface is getting more uniform. Both of the factors are good for arc diffusion. However once there is intense vacuum arc, as in high current mode, the anode will be melted seriously, and plenty of metal vapors will be evaporated when current passes zero. Meanwhile, hot particles will be emitted continuously from the overheated area. Thus the dielectric recovery strength is weakened greatly and the vacuum arc will be reignited and broken down easily.

3.3 Vacuum Arc Influenced by Material of Contact

This section is divided into two groups of experiments to compare the characteristics of VF vacuum arc with the same contact diameter but different materials.

3.3.1 Arc Characteristics in Cu and Cucr50 Contacts

Arc images for Cu and CuCr50 contacts at the moment of peak current are displayed in Fig. 3.5. When the peak current is 20 kA, the Cu contact fails to interrupt vacuum arc. When the current peak is 22.5 kA, the CuCr50 contact cannot interrupt the vacuum arc either.

By analyzing the arc images, we found that as following: As the current increases, the diameter of the vacuum arc increases. When at the same current, arc combustion, arc brightness, and arc energy concentration area are stronger in Cu contact than that in CuCr50 contact. When the current is too high to be interrupted, the vacuum arc is too bright and spills over the edge of the contact. In the Cu contact, the failed interruption is are 20 and 22.5 kA, while in the CuCr50 contact, the failed interruption is 22.5 kA. It is marked by underline in Fig. 3.5. For VF vacuum arc at 360 Hz in shortgap, concentrated state is not observed.

When the VF vacuum arc is interrupted, it is unlikely that there is a violent anode spot, but the surface will melt and evaporate. For adding element Cr, the melting point of CuCr50 alloy is higher than that of Cu, which will reduce the ablation of contact surface. There are obvious melting regions on the surface of the two contacts after the experiment. Some small and shallow corrosion pits are

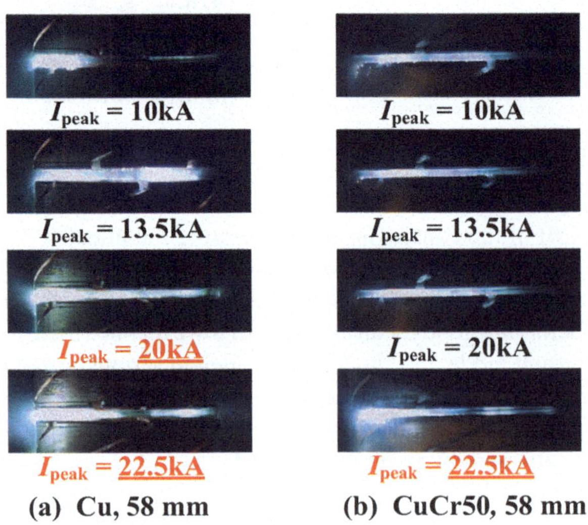

Fig. 3.5 Arc images at peak current of Cu and CuCr50 contacts with diameter of 58 mm (360 Hz)

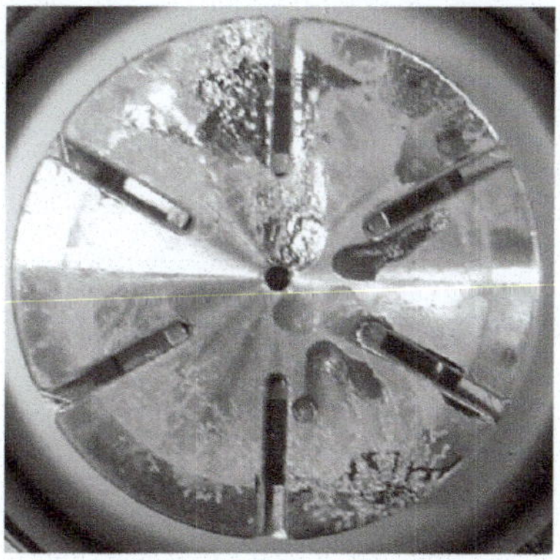

(a) Anode surface of Cu contact

(b) Anode surface of Cucr50 contact

Fig. 3.6 Surface ablation for Cu and CuCr50 contacts

distributed unevenly. There are metal protrusions at the boundary of the melting area, however, with no large amount of ablation in CuCr50. Compared to the CuCr50 contact, the melting phenomena are more obvious and serious in the Cu contact. After repeated experiments of different current levels, the surface ablation of Cu and CuCr50 contacts are shown in Fig. 3.6 respectively.

Voltage of vacuum arc in Cu and CuCr50 contact in low current mode is shown in Fig. 3.7. The arc voltage of CuCr50 contact is significantly lower than that of Cu contact, thus the arc energy of Cu contact is higher when the current is the same, leading to brighter arc in the image. When the peak current is 10 kA, voltages in both contacts are flat. The maximum arc voltage of CuCr50 contact is about 4 V lower than that of Cu contact. When the peak current is 13.5 kA, the arc voltage of CuCr50 contact rises slightly but remains flat. However, there is a spike in the arc voltage of Cu contact, with maximum value of 52.2 V, which is 16 V more than that in the CuCr50 contact. The reason why the CuCr50 contact arc voltage is lower because saturation vapor pressure of Cr is higher than that of Cu, and the plasma density in the arc gap is correspondingly high, leading to small arc resistance. Thus the arc voltage is lower.

Arc voltage in Cu and CuCr50 contact in high current mode is shown in Fig. 3.8. The arc voltage of both contacts increases compared to low current mode. When the peak current is 20 kA, the Cu contact cannot interrupt the VF vacuum arc. As the current peak is 22.5 kA, the CuCr50 contact fails to interrupt either. If the interruption is failed, the oscillation in arc voltage would occur. For a lower melting point, Cu is much easier to be melted than CuCr50. Sequentially, vapors will be evaporated severely and metal droplets will be emitted in large quantities, which makes the broken down and reignited VF vacuum arc easily. It can be seen that the dielectric strength and breaking capacity of CuCr50 contacts are better than that of Cu contacts.

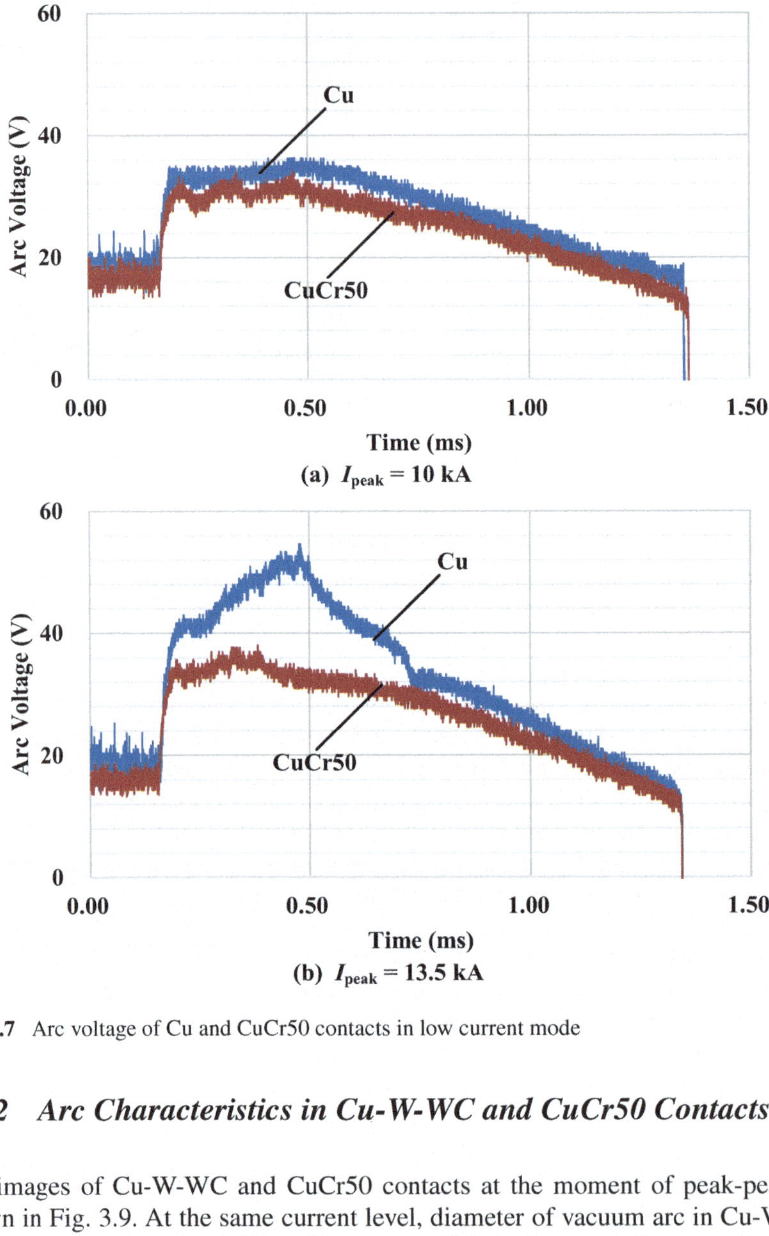

Fig. 3.7 Arc voltage of Cu and CuCr50 contacts in low current mode

3.3.2 Arc Characteristics in Cu-W-WC and CuCr50 Contacts

Arc images of Cu-W-WC and CuCr50 contacts at the moment of peak-peak are shown in Fig. 3.9. At the same current level, diameter of vacuum arc in Cu-W-WC contact is larger than that in CuCr50 contact. With the increase of current, the arc is getting thicker and there are macroscopic droplets emitting from the contact. When the peak current is up to 18.5 kA, current is failed to be interrupted in the Cu-W-WC contact. Simultaneously, the arc is extremely bright and filled the gap between contacts fully. Vacuum arc is too active in the gap to be out of control by the AMF.

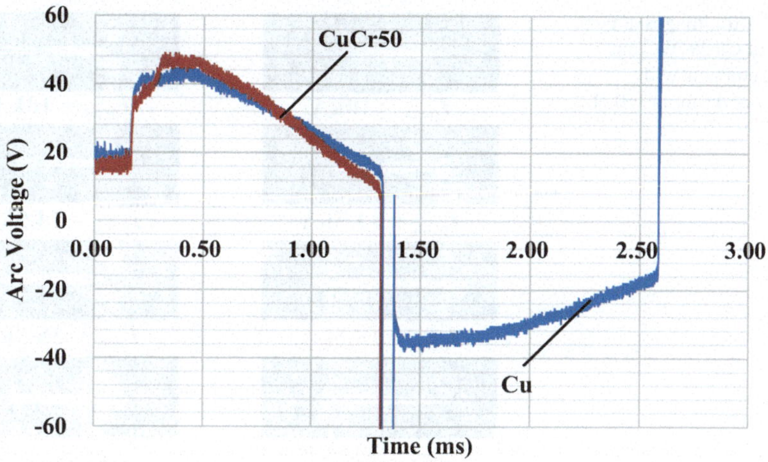

(a) $I_{peak} = 20$ kA

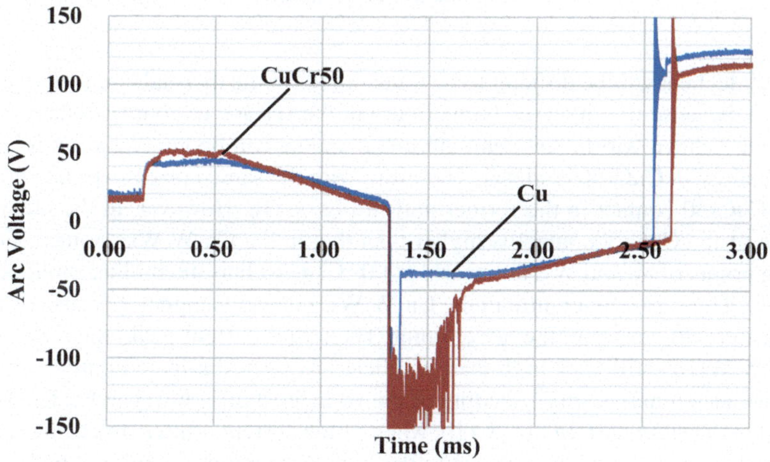

(b) $I_{peak} = 22.5$ kA

Fig. 3.8 Arc voltage of Cu and CuCr50 contacts in high current mode

When the peak current is up to 22.5 kA, current is failed to be interrupted in the CuCr50 contact either. Meanwhile, the arc moves to the side of the contact.

Waveforms of arc voltage in Cu-W-WC and CuCr50 contact with 41 mm diameter are shown in Figs. 3.10 and 3.11. When the peak current is 10 kA, vacuum arcs in both contacts are stable. When the current peak value is 13.5 kA, there is voltage spike in the two contacts, and the maximum value in Cu-W-WC contact is 56.5 V, which is greater than the maximum value of 50.7 V in CuCr50 contact. When the peak current is 18.5 kA, the Cu-W-WC contact cannot interrupt the vacuum arc.

Compared to W and WC, as the boiling point of Cr is low and the vapors pressure of Cr is high, plenty of metal vapors could be generated at lower temperature.

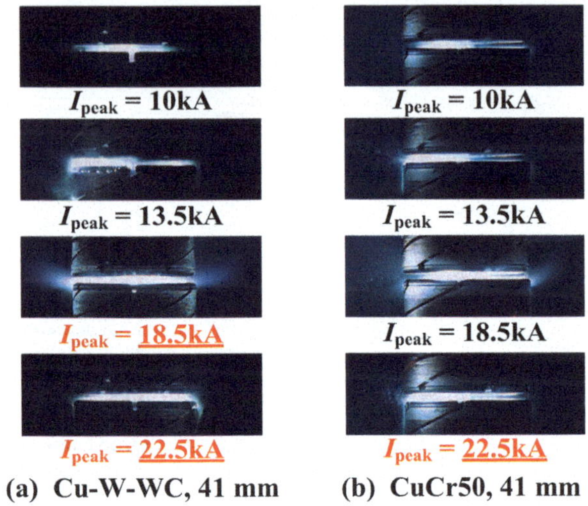

Fig. 3.9 Arc images at peak current of Cu-W-WC and CuCr50 contacts with diameter of 41 mm (360 Hz)

Meanwhile, heat will be difficult to be transmitted for the thermal conductivity of Cr is worse, because of which the temperature in the surface of contact is easy to rise. Due to the above two factors, enough metal vapors can be generated by relatively lower energy in CuCr50 contacts. Therefore with the same current, voltage is lower in the CuCr50 contact in the former half in Fig. 3.11a. However, in the latter half, voltage of the CuCr50contact is higher than that in the Cu-W-WC contact. For the boiling points of W and WC are nearly 6000 °C, as well as the boiling point of Cr is 2672 °C. Thus arc temperature in the Cu-W-WC contact is higher, because of which the plasma will denser in the arc column. Then after current peak, the temperature of Cu-W-WC is still very high even the heat has been dissipated partly. Thus the arc resistance remains small, leading to the voltage value is lower in the Cu-W-WC contact in the latter half. In Fig. 3.11b, when a high current is interrupted, the residual plasma density is too high, and due to the high temperature, there will be hot electron emission. Arc reignition is easy to happen even if the vacuum arc has been extinguished, which is the disadvantage of adding W.

3.4 Analysis of Interruption Characteristics in Different Contacts

In this chapter, four kinds of vacuum circuit breakers are used in the experiment. When the vacuum arc cannot be interrupted, the current peak value is shown in Table 3.3.

It can be seen from the results that, for the contact with the same material, the larger the diameter of contact is, the higher the interruption current is. And the maximum interrupted current represents the breaking or breaking capacity of the

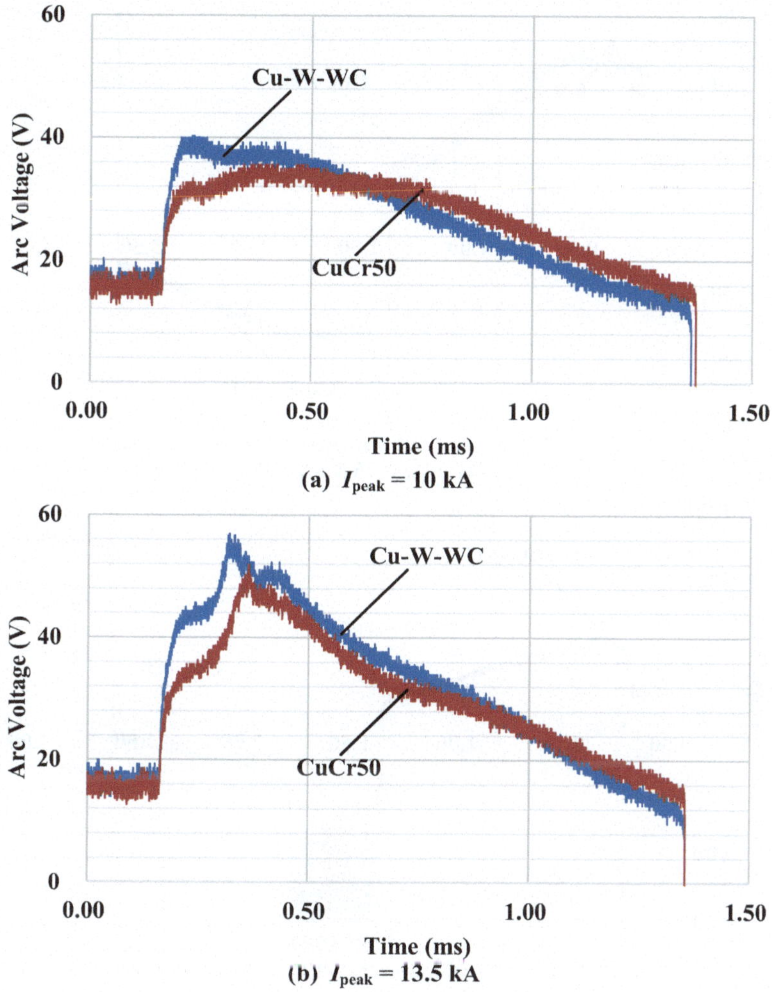

Fig. 3.10 Arc voltage of Cu-W-WC and CuCr50 contacts in low current mode

contact. As shown in Table 3.3, the breaking capacity of CuCr50 is stronger than that of Cu-W-WC and Cu. There is low melting point for Cu which is easy to melt. When applying Cu as the material of contact, evaporation and sputtering may lead to local melting and fusion welding. Thus the metal evaporation density is high, which will cause breakdown and reignition post-arc easily, as well as the breaking capacity is limited. If adding Cr into Cu, the anti-fusion welding ability can be enhanced, and there is also high breakdown strength, lower arc burn loss rate, and low probability of reignition for CuCr50.

Both W and WC are with higher melting point. When Cu-W-WC alloy is used as the material of contact, there will be excellent properties such as strong mechanical

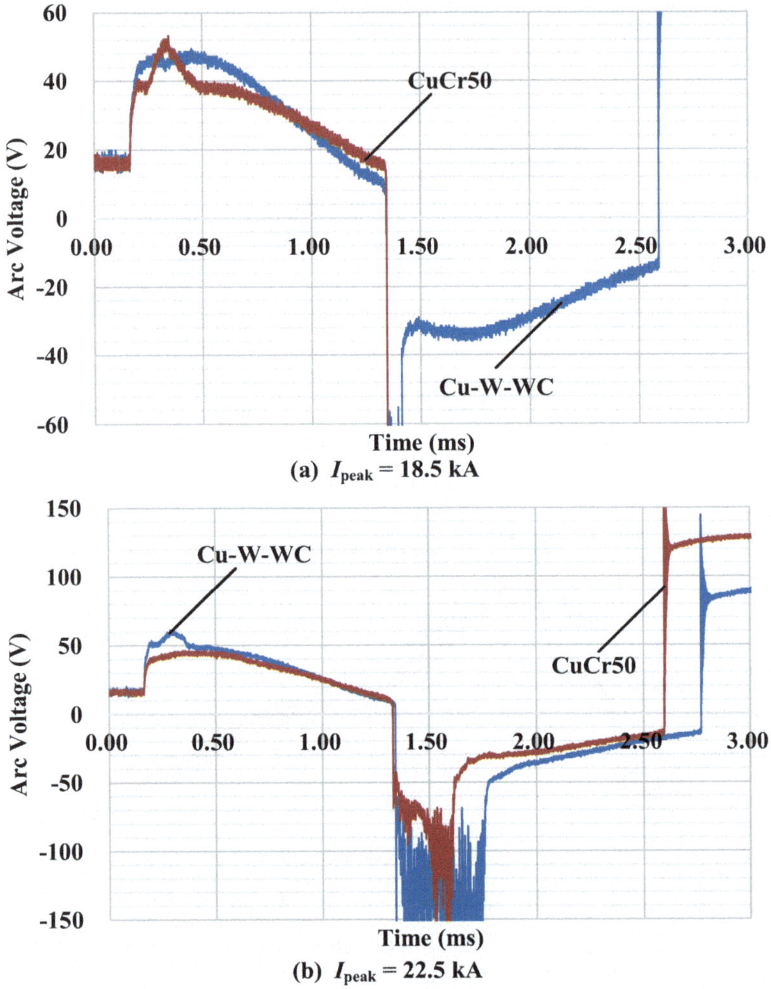

Fig. 3.11 Arc voltage of Cu-W-WC and CuCr50 contacts with diameter of 41 mm in high current mode

Table 3.3 Maximum interrupted current of each vacuum circuit (peak value)

No.	Contact diameter (mm)	Contact material	Maximum interrupted peak current (kA)
1	58	CuCr50	22.5
2	41	CuCr50	20
3	58	Cu	20
4	41	Cu-W-WC	18.5

resistance and light fusion welding tendency. However, the vacuum arc in Cu-W-WC contacts has a high temperature, so there will be more hot electrons emitted from the contact after current zero, leading to high-density residual plasma in the chamber, and making the arc reignite easily.

The peak arc voltages of the four contacts are shown in Table 3.4, and the voltages marked by the underline indicate failed current interruption in the experiment. The curve for the U_{peak} (peak arc voltage between contacts) and the I_{peak} (peak current) is shown in Fig. 3.12.

From Fig. 3.12 we can conclude that: When at low current mode (10 and 13.5 kA), all the four contacts can interrupt current successfully, and U_{peak} increases with the increase of I_{peak}. For the same current, the highest voltage in U_{peak} is in the Cu-W-WC contact with a diameter of 41 mm, and the lowest voltage in U_{peak} is in the CuCr50 contact with a diameter of 58 mm. As the current increases to the value which cannot be interrupted successfully, U_{peak} will drops and show a minimum value in the U_{peak}-I_{peak} curve. After that, when the current continues to increase,

Table 3.4 Peak arc voltage of the four contacts at different currents

Peak current (kA)	Peak voltage of each contact (V)			
	58 mm CuCr50	41 mm CuCr50	58 mm Cu	41 mm Cu-W-WC
10	31.9	36.2	36.5	40.6
13.5	36.2	50.7	52.2	56.5
18.5	44.9	52.2	57.6	47.8
20	48.5	43.5	43.5	53.7
22.5	47.8	46.4	45.2	59.4

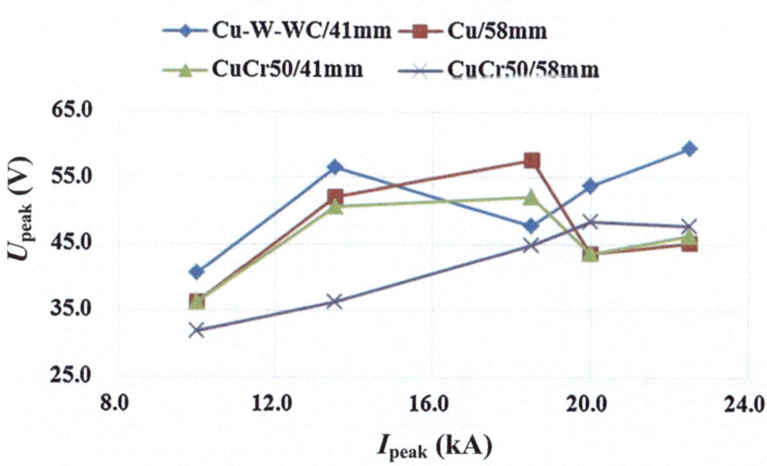

Fig. 3.12 Relationship between peak arc voltage and peak current of each contact

U_{peak} also increases. At the same current, there is still the highest voltage in U_{peak} in the Cu-W-WC contact with a diameter of 41 mm.

This experimental phenomenon can be explained as follows: When in a low current mode, U_{peak} is proportional to the effective value of current for the vacuum arc is a positive resistance. With the increase of current, the plasma density tends to be much higher. If the plasma density grows rapidly, thus the conductivity of vacuum arc will be very good when the arc fills fully in the contact gap, making the current interrupted fail. After that moment, both the resistance of vacuum arc and the arc voltage descend significantly. If the current increases further, the resistance of vacuum arc will not change a lot, because the plasma density has been saturated. U_{peak} is proportional to the current value again for the effect of current plays a major role.

References

1. Yuan Jiang, Jianwen Wu. Effects of Contact Materials and Diameters on Characteristics of Intermediate-frequency Vacuum Arc in Axial Magnetic Field [J]. Proceedings of the CSEE, 2015, 35(20): 5367–5374.
2. Jimei Wang. Contact material and manufacturing technology of vacuum switch [M]. Xi'an: Research center of vacuum arc theory, Xi'an Jiaotong University, 1995.
3. Peter Frey, Norbert Klink, Roland Michal, et al. Metallurgical aspects of contact materials for vacuum switching devices [J]. IEEE Transactions on Plasma Science, 1989, 17(5): 734–740.

Chapter 4
Frequency Characteristics of Variable Frequency Vacuum Arc

As is known, in the aviation power supply system, the increase of current frequency will lead to the increase in the change rate, d*i*/d*t*, when current crossing zero, and accelerate the rise rate of recovery voltage, d*u*/d*t*, and the eddy current effect will be more obvious. When a short circuit fault occurs, the instantaneous current peak value can reach more than 20 kA, and the interruption of the arc will become very difficult under this circumstance [1, 2]. In this chapter, the influence of frequency changes (360–850 Hz) on the characteristics of VF vacuum arc will be discussed.

4.1 The Effect of Frequency on Vacuum Arc in AMF Contact

4.1.1 Interruption-Frequency Characteristics in AMF

The breaking capacity is one of the crucial technical indicators of the vacuum switch. We will not only look into the breaking capacity of the vacuum interrupter at different frequencies but also analyze the main factors affecting its breaking performance. In the section, we use the oscillation circuit to change the frequency of the power supply and study the limit breaking current capacity of the cup-shaped longitudinal magnetic vacuum interrupter. Vacuum circuit breaker (VCB) of AMF type with contact materials of CuCr50 and diameter of 41 mm is tested in this section, with a contact distance of 3 mm. By changing the current frequency, we obtain the current peak value in the first half-wave for successful and failed interruption, as shown in Table 4.1, represented by I_{limit}. To ensure accuracy, each test is performed more than 15 times. It can be seen that with the increase of frequency, its breaking capacity decreases.

Y. Jiang and Q. Li, *Vacuum Circuit Breaker for Aviation Variable Frequency Power System*, Power Systems, https://doi.org/10.1007/978-981-33-4781-6_4

Table 4.1 Limit current for interruption at different frequency (AMF, CuCr50, 41 mm)

f (Hz)	$T = 1/f$ (ms)	I_{limit} (kA)
360	2.78	22.4
500	2.00	16.5
600	1.67	15.5
680	1.47	15.0
850	1.18	14.0

The interruption current is 50 kA/40 kA for the experimental VCB, with the conditions of 380 V/660 V at a power frequency of 50 Hz/60 Hz. However, according to these results, the breaking capacity decreases significantly in the VF experiment. In the former power frequency vacuum arc investigation, Voshall believed that the current breaking capacity was inversely proportional to the frequency [3].

In order to obtain the relationship between the limit interruption current and frequency, according to the experimental results, taking the frequency of current (Hz) as the horizontal ordinate and the limit current value (kA) as the vertical ordinate, the curve is drawn as shown in Fig. 4.1.

To facilitate linearization analysis, the frequency is taken as the reciprocal, and the data is further processed within the range of the current frequency from 360 to 850 Hz: taken the square of the intermediate frequency current period as the abscissa, and the peak value of the limit breaking current as the ordinate, an approximate linear curve can be drawn as shown in Fig. 4.2.

It can be seen that the limit breaking current value and the square of the period are approximately linear. Furthermore, the linearity is better. Using the least-squares method, the relational expression is:

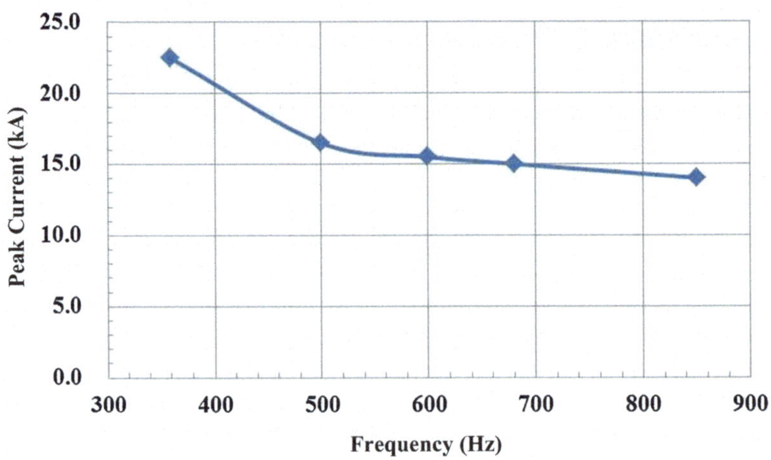

Fig. 4.1 Limit breaking capacity at different frequencies (AMF, CuCr50, 41 mm)

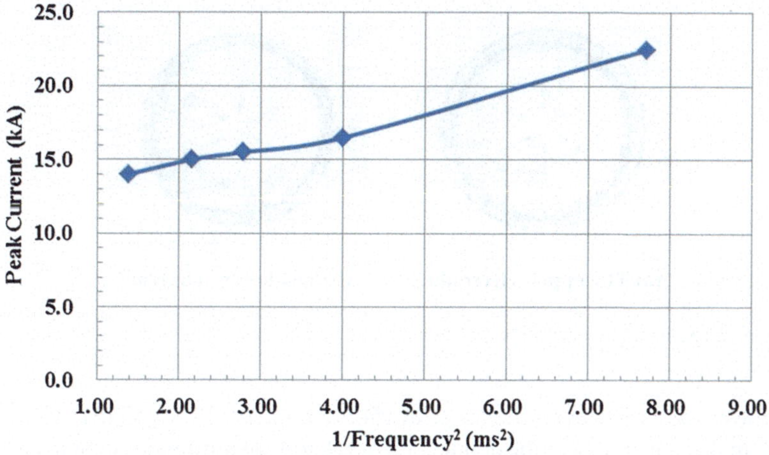

Fig. 4.2 Relationship of limit breaking current and period squared (AMF, CuCr50, 41 mm)

$$I_{\text{peak-limit}} = \frac{a}{f^2} + b = aT^2 + b \tag{4.1}$$

where $a = 1.34$ kA/ms^2, and $b = 11.86$ kA, and the curve fitting correlation coefficient, $R = 0.9833$.

The extinguishment of the vacuum arc relies on the recovery process of the remaining plasma parameters between the arc gaps after the current crossing zero. Factors such as the change rate at the time of current zero-crossing di/dt, the rise rate of recovery voltage, the hysteresis of the AMF, and the thermal state of the anode contact will all limit the current breaking capacity of VCB. The above-mentioned experimental research and data fitting found that the limiting current breaking capacity of the AMF interrupter is inversely proportional to the square of the frequency. Since the recovery voltage is basically the same during the experiment, the dI/dt and the hysteresis of the AMF in the case of intermediate frequency are of great importance to the breaking of VF vacuum arc [4].

But at the same time, compared with the power frequency current of the same current value, the arcing time of the VF vacuum arc is shortened and the arc energy is reduced. Therefore, the ablation of the anode contact is reduced and the possibility resulting in violent anode activity like under power frequency conditions is reduced. This inference can be verified from the contact ablation that reflects the thermal state of the contact during the arcing process. As shown in Fig. 4.3, the maximum current peak of the multiple intermediate frequency experiments climbs to 22.5 kA in CuCr50 contacts. Among them, in the successful breaking experiment, the upper contact is used as the anode, and the lower contact is used as the cathode; when the breaking fails, the current polarity is reversed after reignition whereas the upper contact is used as the cathode, and the lower is used as the anode. So overall, the upper contact operates longer as the anode.

(a) The upper electrode (b) The lower electrode

Fig. 4.3 CuCr50 contacts after VF experiments

In power frequency experiments (CuCr50/41 mm, 10–17.5 kA) [5], the ablation appears in the center area of the contact, whose degree of the anode is obviously more serious than that of the cathode and. For the anode, the plasma in the arc column area has a large pressure when it reaches the anode. Under the action of the electromagnetic force, the arc will shrink near the anode, especially at high current mode. The anode may even cause arc accumulation. The degree of ablation is obviously serious and appears in the central area; due to the different erosion mechanism, the ablation of the cathode surface comes from the effect of the cathode spots. The number of cathode spots is large and evenly distributed on the contact surface, so the ablation is relatively uniform. The more serious cathodic ablation at 17.5 kA is the result of arc reignition, as displayed in Fig. 4.4.

From the comparison of the results, it can be seen that the surface of the contact after the intermediate frequency experiment is unevenly distributed with melting marks, and the ablation of the anode is also more evident than that of the cathode, but there is no severe large-area ablation. It can be drawn that VF vacuum arc did not form an active anode spot mode or concentrated arc mode.

The reasons for the above experimental results may be explained from two perspectives. One is because the burning time of the VF arc is shorter than the power frequency, the power frequency half-wave time is 10 ms whereas the 360 Hz current half-wave time is only 1.4 ms. The difference between the two is as much as 7 times. The higher the frequency, the shorter the half-wave time is. At the same current level, the power frequency arc injected more energy into the contacts. On the other hand, due to the limited gap in the experiment (3 mm), the arc contraction near the anode was not obvious, no active anode spot or concentrated mode were detected.

4.1.2 Voltage-Frequency Characteristics in AMF

Figure 4.5 demonstrates the relationship between time and voltage of vacuum arc as well as time and current under different situations.

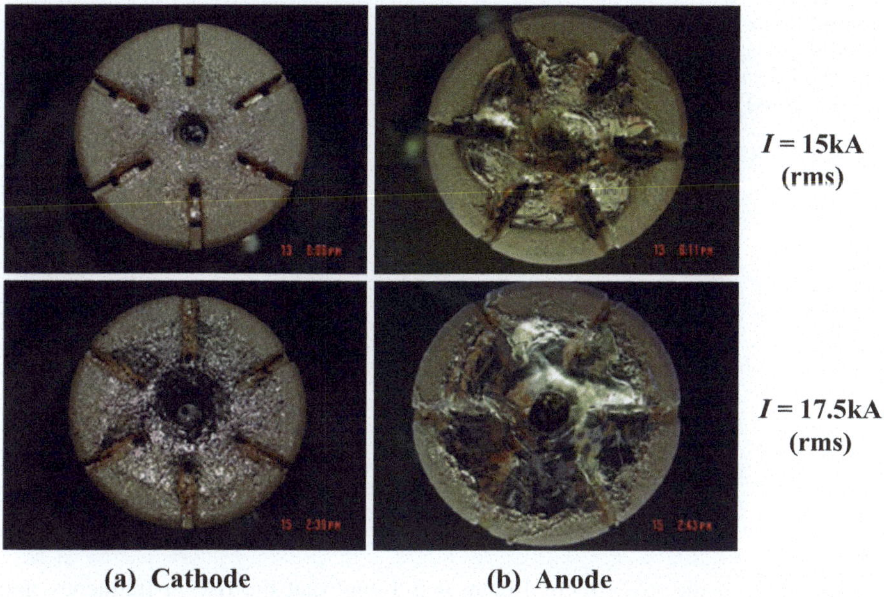

Fig. 4.4 CuCr50 contacts after power frequency experiments

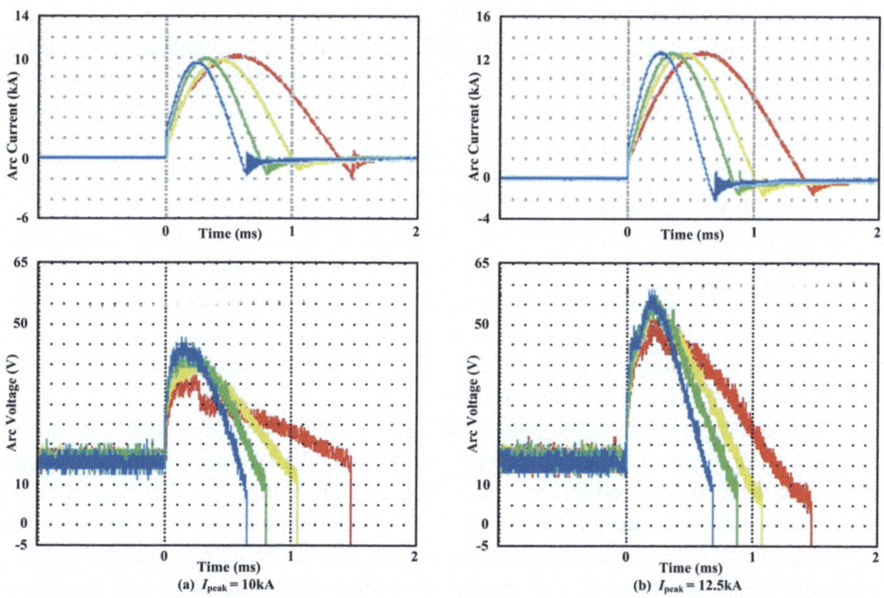

Fig. 4.5 Vacuum arc current and voltage in different frequencies ($I_{\text{peak}} = 10$ and 12.5 kA)

Table 4.2 Peak value and time of arc voltage at different frequencies (I_{peak} = 10 and 12.5 kA)

I_{peak} (kA)	f (Hz)	$T = 1/f$ (ms)	U_{peak} (V)
10	360	2.78	36.0
	500	2.00	41.3
	680	1.47	43.3
	850	1.18	44.0
12.5	360	2.78	50.0
	500	2.00	56.0
	600	1.67	57.3
	780	1.28	58.7
14.4	360	2.78	61.4
	500	2.00	65.3
	600	1.67	68.3
	680	1.47	69.0
	850	1.18	70.3

According to the experimental data, it is found that the rise in frequency and current leads to an increase in voltage. First, due to the thermal inertia of plasma, the ionization is inadequate. The arc resistivity basically remains unchanged when the current is increasing. Second, the arc path curved when it was burning at the lateral of contacts, which can be seen as lengthened. As a result, the arc resistance increases.

It can be seen from Table 4.2 that as the current frequency goes up, the voltage peak value also goes up; meanwhile, the time when the peak value appears advanced as well. Regard moment when the VF current is introduced at zero: when the current is 10 kA, the voltage peak value increases from 36 V at 360 Hz to 44 V at 850 Hz, and the peak moment is advanced from 270.0 to 137.6 μs; when the current is 12.5 kA, the voltage peak increases from 50 V at 360 Hz to 58.7 V at 780 Hz, the peak moment is advanced from 248.2 to 199.6 μs. The increase of the voltage peak value and the advancement of the peak time indicates that the voltage rise rate du/dt grows with the increase of frequency. The arc voltage gradually decreases after the peak value, the higher the frequency, the faster the arc voltage drops, and the instantaneous value before zero crossings has little difference.

Figure 4.6 displays the relationship between the peak arc voltage and the current cycle when the breaking current is 10 kA, 12.5 kA, and 14.4 kA respectively. It can be observed that if the frequency is the same, the arc voltage peak value goes up when the current peak value is large; while if the current peak value is the same, the arc voltage peak value increases with the growth of the frequency and the voltage peak value and the current period are approximately linear, furthermore, the slopes of the three sets of curves are basically consistent with the changing trends.

If the current frequency f is regarded as the independent variables, and the voltage peak U_{peak} as the dependent variable, the following function can be obtained:

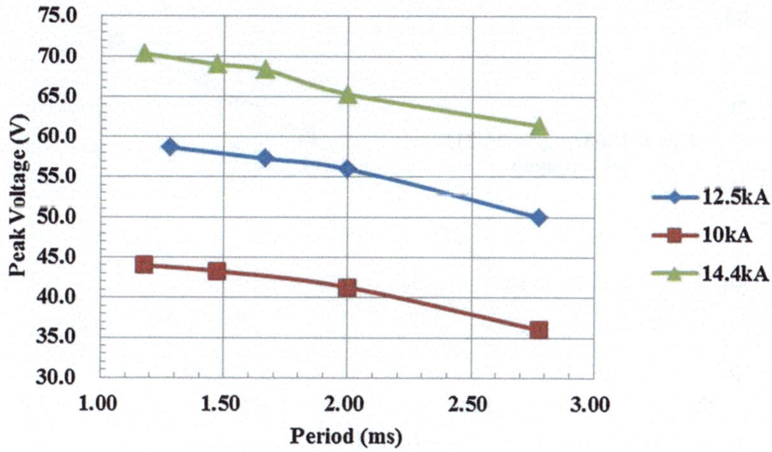

Fig. 4.6 Peak value of arc voltage in different frequencies

Table 4.3 K_f and U_I calculated by fitting

I_{peak} (kA)	U_I (V)	K_f (V/s)
10	50.540	5058.9
12.5	66.928	5916.2
14.4	77.331	5758.7

$$U_{peak} = U_I - K_f \cdot T = U_I - \frac{K_f}{f} \tag{4.2}$$

U_I—reflects the influence of the peak current on the arc voltage (unit V),
K_f—represents the coefficient (unit V/s) related to current frequency (period).

By fitting the experimental data, the calculated values are shown in Table 4.3.

Judging from the calculation results, K_f does not change much while U_I changes incrementally. The reasons for this phenomenon are as follows: in the linear relationship, K_f is used as the coefficient of frequency f, which itself should not be linear with f. In this experiment, only one vacuum interrupter is used and the distance is constant, so K_f should be related to the structural parameters such as contact structure, contact diameter, and contact distance. U_I reflects the influence of the peak current on the arc voltage. In the study of power frequency vacuum arc, within a certain range of the current, the voltage value of the vacuum arc increases with the increase of the current. Furthermore, a similar conclusion can be drawn in the 400 Hz experiment.

After fitting U_I and I_{peak}, the following equation can be obtained:

$$U_I = 0.2503 I_{peak}^2 + 26.215 = K_I \cdot I_{peak}^2 + U_C \tag{4.3}$$

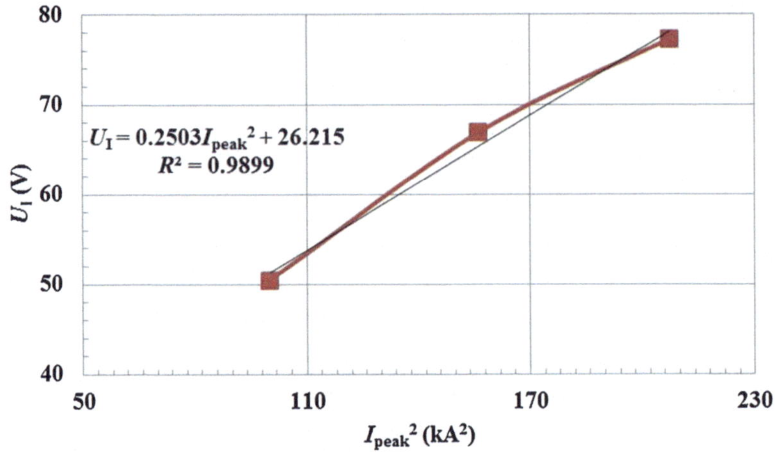

Fig. 4.7 The relationship between U_I and I_{peak}

K_I—the coefficient related to the current peak value (unit V/kA^2).

U_C—voltage constant (unit V).

The curve is shown in Fig. 4.7.

In summary, substitute Eq. (4.3) into Eq. (4.2), and the relationship between the voltage peak, current peak, and period can be expressed as follows:

$$U_{peak} = K_I \cdot I_{peak}^2 - \frac{K_f}{f} + U_C \quad (4.4)$$

According to Eq. (4.4), the arc voltage peak value increases with the increase of frequency and current peak value. Set the current frequency range between 360 and 800 Hz and the peak current between 10 and 20 kA. The function relationship between U_{peak} and f and I_{peak} is drawn in a 3D diagram, as shown in Fig. 4.8.

Whether increasing the current value or increasing the current frequency, both situations will contribute to a rise in the current change rate, di/dt. First, more electrons will be injected into the arc in a unit of time, so more ions will reach the anode. In this way, the arc voltage must be quickly increased to a sufficient amplitude within the same time to generate enough positive ions to establish an electric field and attract the surge of electrons. Second, due to the magnetic field hysteresis of the axial magnetic contact at the beginning of the current, the vacuum arc is in a relatively weak AMF confinement environment, and the positive ions can quickly diffuse to the outer space of the contact, making the cations near the anode scarce, and the arc voltage must also rise quickly to maintain the plasma balance. And as the current frequency increases, the magnetic field will lag further, and the tendency of scarcity will become more serious. Thirdly, the higher the frequency, the more serious the shortage. Electrons are emitted from the cathode. Since the characteristics of the plasma are quasi-neutral, the distance between electrons and ions will not be

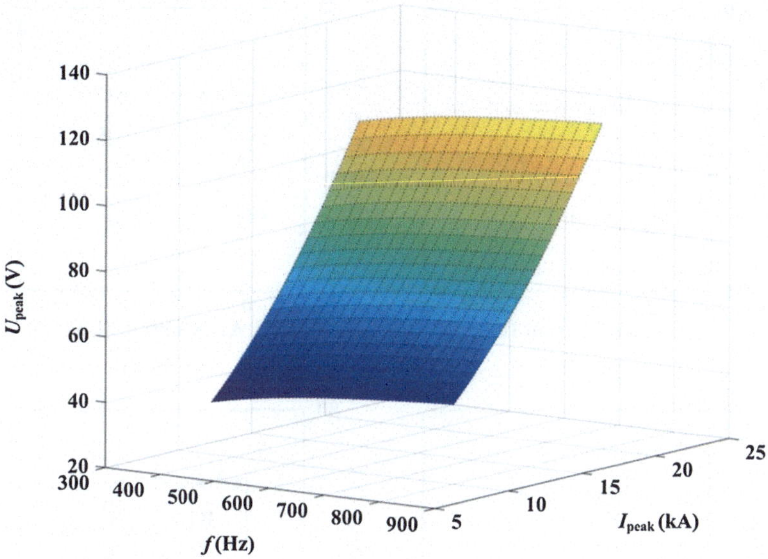

Fig. 4.8 Relationship between U_{peak}, f, and I_{peak}

greater than the Debye length. In the micro-area, the electrons turn around the ions at high speed and ions move slower than electrons. As a result, there is a greater impact on ions than electrons by a lack of magnetic restriction. In conclusion, all these factors lead to an increase of the arc voltage peak value and the voltage rise rate d*u*/d*t*.

4.1.3 Evolution of IF Vacuum Arc in AMF

The dynamic volt-ampere characteristics of the vacuum arc can intuitively reflect the dynamic resistance characteristics of the arc. The research results of the power frequency vacuum arc show that the vacuum arc has positive volt-ampere characteristics, that is, the arc voltage increases with the increase of current. The arc dynamic volt-ampere characteristics can be acquired by obtaining the arc voltage and current waveform, with the arc current as of the horizontal axis and the arc voltage as the vertical axis, and its slope represents the arc dynamic resistance.

Figures 4.9 and 4.10 show the dynamic volt-ampere characteristics of the intermediate frequency vacuum arc at different frequencies when the current peak reaches 10 kA and 12.5 kA, respectively. From the different volt-ampere characteristic curve, the same law can be observed, that is, as the current and voltage change, the arc change process can be divided into the following three stages, as shown in Figs. 4.9 and 4.10. The first stage starts at the introduction point A of intermediate frequency current,

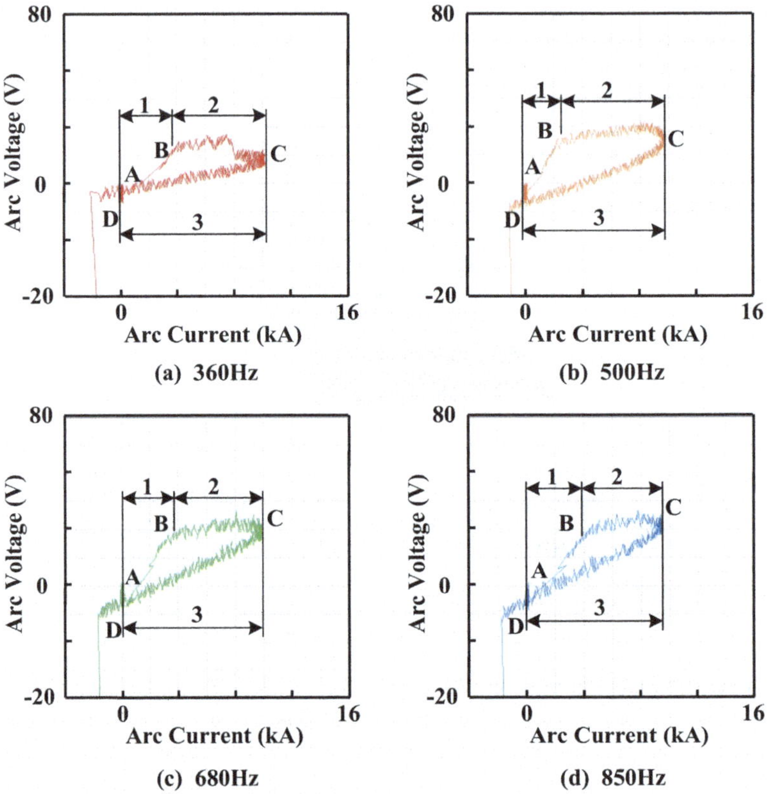

Fig. 4.9 Volt–ampere characteristics of VF vacuum arc (10 kA)

marked by the rapid and smooth rise of the arc voltage, and ends with the slope resistance change B in the dynamic volt-ampere characteristics. At this time, the arc dynamic volt-ampere characteristics present positive resistance characteristics. The second stage starts with B after the end of the free expansion stage and ends at C at the current peak. The third stage starts after the current peak value C and continues to the current zero-crossing point D.

In order to study the evolution of the arc appearance, the volt-ampere characteristic, arc voltage, and current changing curve are analyzed together with the arc image. Taking 360 and 500 Hz as examples, the curves of arc voltage and current with time at the two frequencies and the arc images at certain moments are shown in Figs. 4.11 and 4.12, respectively. Since the sampling rate of the oscilloscope is much greater than the sampling time of the high-speed camera, if we look for the corresponding arc image with an arbitrary point on the arc voltage waveform as a reference, generally we can only find two consecutive images in time. Therefore in Figs. 4.11 and 4.12, for the time of point A, the corresponding arc image time should be t_{A-} and t_{A+}, and the same rule applies for point B, C and D. However, time t_1 and

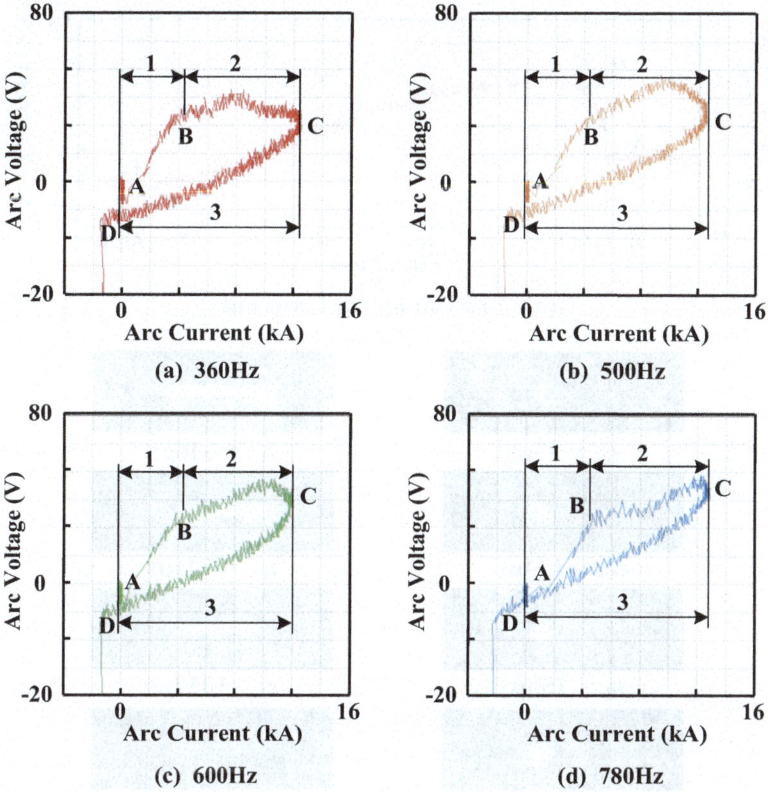

Fig. 4.10 Volt–ampere characteristics of VF vacuum arc (12.5 kA)

t_2 are referred to as the arc image, and the exact corresponding time can be found on the voltage waveform.

Figures 4.11 and 4.12 show the curve of voltage and current of intermediate frequency vacuum arc at 360 and 500 Hz, as well as the arc image in the evolution of arc. The peak value of arc current is 12.5 kA. Since the law of evolution is similar, we take Fig. 4.11 as an example. It can be seen that the evolution of vacuum arc can be divided into three stages. S1 starts at t_A when the AC current injected. For the magnetic field lags behind the current due to the eddy current effect, and the vacuum arc is under the action of a small magnetic field at this time. The arc begins to expand in a circular shape at the position of the DC arc ignition. At this time, the arc voltage rises relatively steadily until it reaches the change point B of the arc dynamic volt-ampere characteristics, and the arc voltage at point B reaches 42.7 V.

When at t_1, the vacuum arc has evolved to S2. This stage continues from t_B to the current peak point t_C, and the magnetic induction intensity between contacts gradually increases at this stage. The arc cathode spot is generated continuously around the twisted arc column, and multiple cathode spots appear around the arc

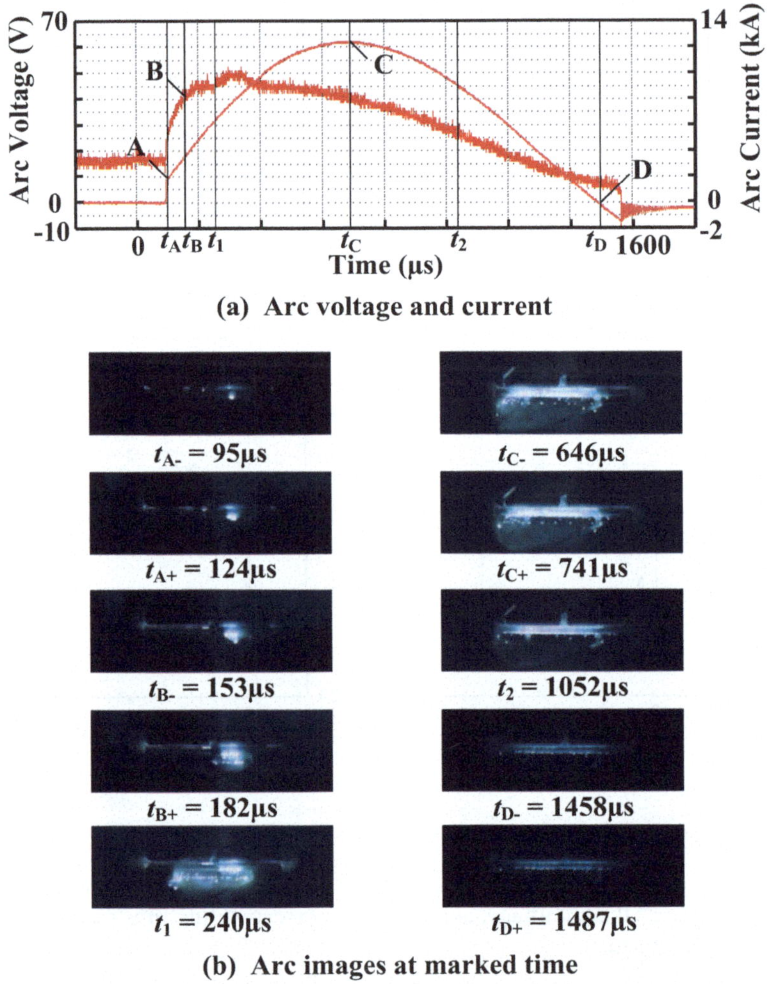

Fig. 4.11 Arc voltage and current waveforms and arc images at 360 Hz (12.5 kA)

column during this process. These cathode spots quickly expand to the surroundings, and the brightness of the arc column rapidly increases. At the same time, the arc voltage fluctuates. The peak value of the arc voltage is 50.0 V, and the arc voltage at the current peak is 41.3 V. This stage is similar to the transient state arc.

From t_C, the arc evolves to S3. Under the constraint of the axial magnetic field, the arc gradually involves into a diffuse state. The cathode spot and arc column stop the rapid expansion trend of the previous stage and occupy the maximum contact surface area in the whole arc evolution process. The arc does not occupy the entire contact surface, and the diffused arc also presents the characteristic of multi-arc column distribution on the cathode surface. The arc column of the arc is separated

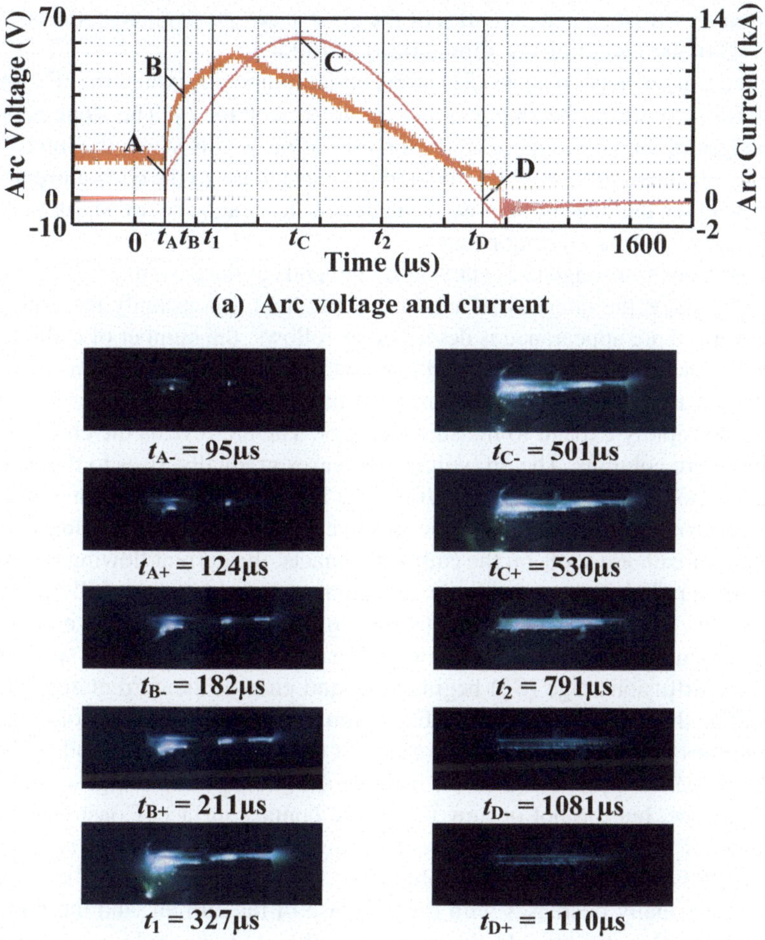

Fig. 4.12 Arc voltage and current waveforms and arc images at 500 Hz (12.5 kA)

into several parts. One part has a bright core, the other part is a scattered cathode spot, and the bright core is not necessarily in the middle of the contact. As the current climb to the peak and gradually starts to drop, both the brightness of arc column and the number of cathode spots show the tendency to decrease, and there is no longer the rapid motion characteristic of S2. This process is maintained until the current reaches zero at t_D. In the meantime, the arc voltage also decreases gradually with the decrease of the current until it arrives at 6 V near the current zero, and then reverses. Due to the hysteresis caused by the eddy current effect, the axial magnetic field reaches the peak after the current peak, and gradually decreases with the decline of arc current. When the current passes zero, the axial magnetic field still retains a certain value.

Combined with the characteristics of the diverse arc shape change, the above S1, S2, and S3 of arc evolution are summarized as follows.

The arc free expansion stage (S1) starts at the moment t_A when the IF current is introduced and ends at the change point in the arc voltage t_B. The slope resistance also changes in the volt-ampere characteristics at t_B. In this stage, the intermediate frequency axial magnetic field lags behind the current due to the eddy current effect, the arc is under the control of a weak magnetic field, and the arc starts to expand freely in a circle at the starting position.

The arc transition stage (S2) starts at t_B and ends at the current peak moment t_C. During this stage, the intensity of axial magnetic field is constantly increasing. The main feature of arc appearance is described as follows: the number of cathode spots goes up around the rotating arc. With the gradual increase in the axial magnetic field intensity, the arc column splits, and a number of cathode spots appear around the arc column and rapidly expand to the surroundings. The arc reveals the characteristics of multiple arc columns. The arc voltage increases gradually, rises to the peak, and then begins to fall off. At the same time, the brightness of the arc increases with the increase of the current value. Due to the expansion of the arc column and the movement of cathode spots on the cathode contacts, the current flowing through the arc gap has a radial component. In the axial magnetic field, the arc will be forced to start to rotate, so it presents a horizontal movement plus a rotating movement mode, making the cathode spots more easily to be separated from the cathode spot clusters.

The arc diffusion stage (S3) begins at t_C and ends at the current zero-crossing time t_D. The diffusion phase of the VF vacuum arc occurs at the peak of the current and continues until the current passes zero. At this stage, the arc cathode spot stops the rapid expansion process, and the cathode spot occupies the largest area on the contact surface, but did not occupy the whole contact surface. Constrained by the axial magnetic field, the arc presents a diffusion state, and the diffused arc may have the characteristics of multiple arc columns. After the current peak, the brightness of the arc gradually decreases with the decrease of the current, and the number of cathode spots also declines. In the meantime, the arc voltage decreases gradually without apparent arc voltage fluctuation until the arc is completely extinguished. Arc volt-ampere characteristics show the characteristics of positive slope resistance, and the slope does not change much when the current decreases from peak value to zero value. Owing to the eddy current effect, the axial magnetic field falls behind the current, reaches a peak value in this stage, and still exists when the current crosses zero.

4.2 The Effect of Frequency on Vacuum Arc in Butt Contact

In the first section of this chapter, the effect of frequency on the current interruption and arc voltage of vacuum arc in AMF contact is studied. As related research, the

Table 4.4 Limit current for interruption at different frequency (butt, CuCr50, 20 mm)

f (Hz)	$T = 1/f$ (ms)	I_{limit} (kA)
360	2.78	14.8
500	2.00	14.3
750	1.33	13
875	1.14	12.4

following section goes on to conduct several experiments on CuCr50 contacts with a diameter of 20 mm. The data listed in Table 4.4 are the limit breaking current peak values of 20 mm flat CuCr50 contacts at different frequencies.

From the data listed above, it can be observed that as the frequency increases, the limit breaking current capacity of the flat contact shows a weakening trend, but the amplitude of the current peak value decrease is not evident. The limit current (kA) and frequency (Hz) of the plate contact are plotted in Fig. 4.13.

The relationship between the two is obtained by fitting the least square method:

$$I_{\text{limit}} = -0.0048f + 16.582 \tag{4.5}$$

Given the fact that the butt contact does not generate AMF, the diffusion of plasma will not be constrained due to the hysteresis of magnetic field when the current crosses zero. Theoretically, this is a favorable factor for breaking. Therefore, as mentioned above, the rate of change at the time of current zero-crossing will be the most crucial factor affecting the current breaking of the flat contact. When the frequency is low, di/dt is small, and the charge of ions injected into the arc gap and electrons can be kept in balance when the current gradually decreases. However, when the frequency is high, di/dt is high, and the electrons injected into the arc gap will decrease faster, moreover, the speed of the electrons will decrease even faster. Then the remaining

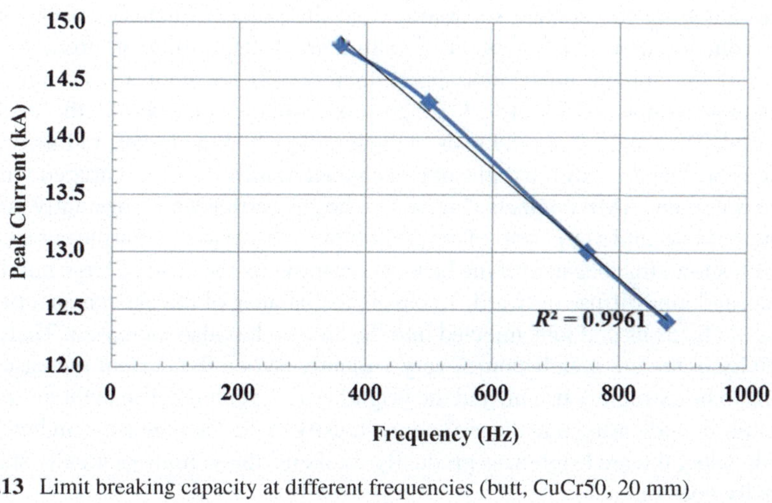

Fig. 4.13 Limit breaking capacity at different frequencies (butt, CuCr50, 20 mm)

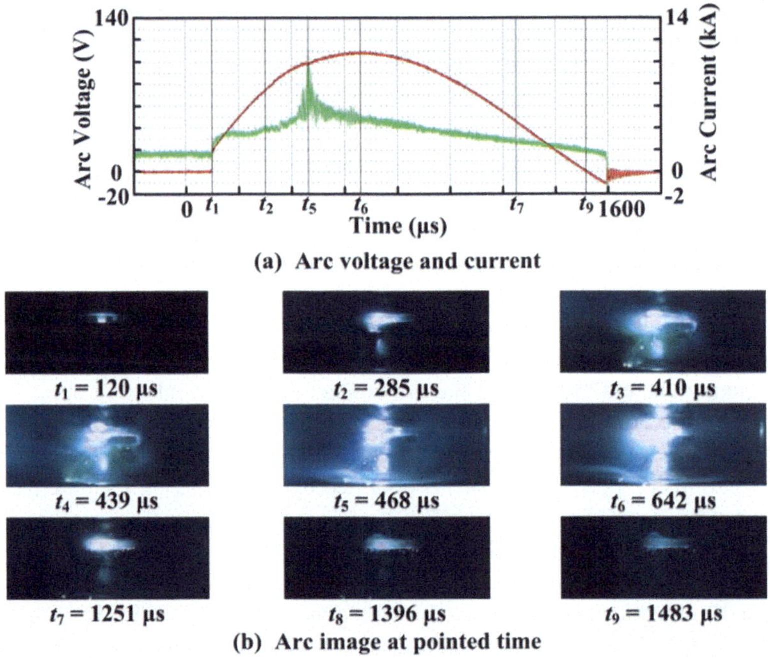

Fig. 4.14 Arc voltage, current, and image in the butt contacts at 360 Hz and 11 kA peak current

ionic current in the arc gap will be larger, which consequently increases the difficulty to interrupt the arc.

The arc voltage, current, and images of arc evolution in the butt contacts at a frequency of 360 Hz and a peak current of 11 kA are shown in Fig. 4.14. When AC current is injected at t_1, the arc appears as a bridge column. At t_2, the arc begins to break out from the contact separation, and the voltage fluctuates according to the arc voltage curve. At this point, a visible arc voltage noise is observed. The duration of the voltage noise was approximately 420 μs, with a maximum peak value of approximately 75 V at t_5. During the arc voltage noise period, the arc shape is distorted. The smallest arc radius is in the middle while the largest radius is at the cathode root. The arc exhibited plasma jet characteristics [6, 7]. Compared with the vacuum arc in the AMF contacts (Fig. 4.11), bright particles are constantly emitted from the cathode, and a large part of the arc column area exists outside the contact gap. The result shows that because of the lack of constraint by the AMF, a large number of electrons and ions diffuse outward. To ensure the balance of the plasma quantity, the number of electrons and ions injected into the arc gap has also increased. Therefore, the activity in the arc area becomes very violently. When the current increases, the arc turns to be extremely bright, and the brightness is even higher than the frequency shown before, indicating that the arc energy density in the flat contact is higher. After the peak value, the arc brightness gradually weakens, the volume gradually shrinks, and finally goes out at t_9.

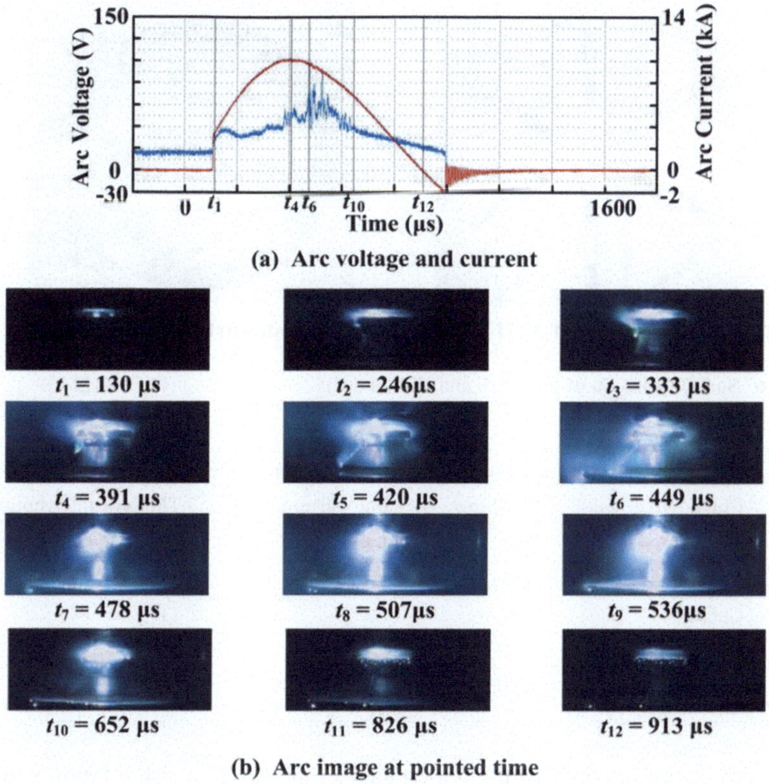

Fig. 4.15 Arc voltage, current, and image in butt contacts at 650 Hz and 10 kA peak current

Similar phenomena can also be observed in Fig. 4.15, where the arc voltage, current, and images of the arc evolution in the butt contacts are set with a frequency of 650 Hz and a peak current of 10 kA. During the arcing period, the arc is constricted, distorted, and moved towards the side of the contact surface accompanied by visible voltage noise. This result proves that under the circumstance of AMF constraint shortage, a huge quantity of electrons and ions were able to escape, and a plasma jet was formed. Meanwhile, as the arc pressure is higher in butt contacts (according to the result in 4.3.3), the vacuum arc would spread outside severely. Consequently, the activity in the area of the arc was very intense. The arc also exhibited plasma jet characteristics.

As explained in the above analysis of the arc images and arc voltage, the arc voltage is higher and the arc appearance becomes more irregular in the butt contacts than those in the AMF contacts under the same current level and experimental conditions. The voltage noise is caused by the arc instability. Images of the surface ablation of the AMF and butt contacts in Fig. 4.16 shows that the ablation of the butt contact is more serious in comparison with the other one because of the more violent arc movement and higher arc energy.

(a) Anode surface of AMF contact **(b) Anode surface of butt contact**

Fig. 4.16 Surface ablation of the AMF and butt contacts

Table 4.5 The limit current interrupted and the times of interruption of experiment contact

No.	Contact material	Structure	Limit current (kA)	Times of interruption
1	CuCr50	Butt	14.4	7
2	CuCr50	Butt	14.4	8
3	CuCr50	Butt	11.3	11
4	CuCr50	AMF	22.4	20
5	CuCr50	AMF	22.4	28

The direct consequence of the flat contact's inability to resist ablation is that its breaking times are less than that of the AMF contact. Evidence can be found in the limit current interrupted and the times of interruption of experiment contact in Table 4.5. The number of breaking times in the table refers to the fact that after several experiments, the breaking capacity of the contact is greatly reduced and the inner wall of the glass shield is covered by a large area of molten metal and the arc image cannot be captured. The contact can no longer be used for valuable experiments. It can be said that the number of breaks of the flat contact is less than that of the AMF contact.

4.3 Analysis of Arc Burning Characteristics of VF Vacuum Arc

4.3.1 Stabilization by AMF

The magnetic induction intensity of the arc area and the middle plane of the distance at the time of the current peak is shown in Figs. 4.17 and 4.18, indicating that the magnetic field is not uniformly distributed. It becomes weak near the center and

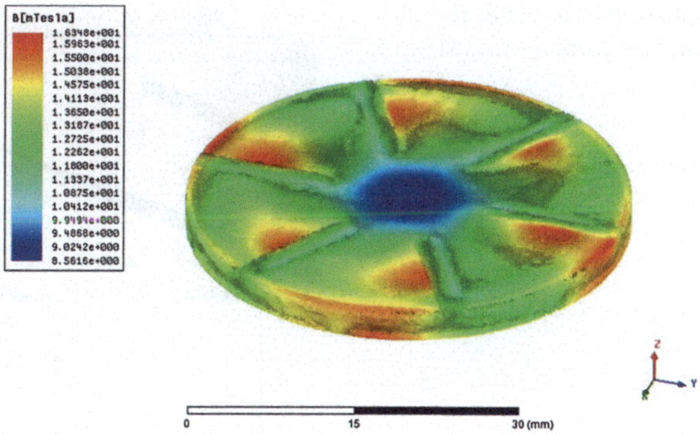

Fig. 4.17 Magnetic flux density in the arc column when the current is at peak

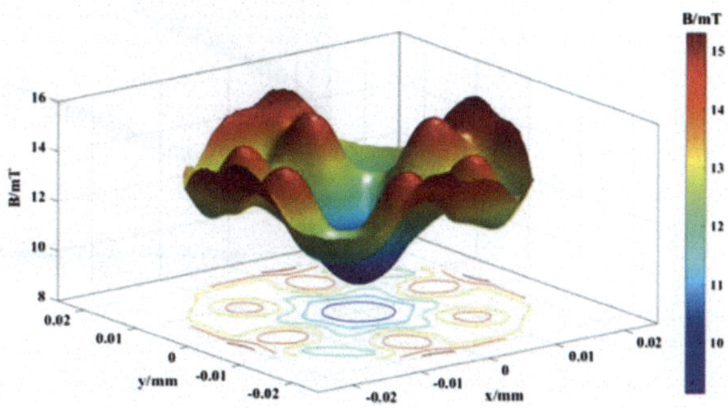

Fig. 4.18 Magnetic flux density in the mid-separation plane when the current is at peak

grows stronger in the surrounding area. The maximum magnetic flux density is approximately 16.3 mT.

During the arcing period, the arc plasma is subjected to the dual effects of the AMF and electric field. If the induction field generated by the motion of the electrons and ions is negligible, there is no collision between plasmas, and the influence of gravity is ignored, the motion equation of the charged particles can be expressed as follows:

$$m\frac{\mathrm{d}\boldsymbol{v}}{\mathrm{d}t} = q\boldsymbol{v} \times \boldsymbol{B} + q\boldsymbol{E} \tag{4.6}$$

The electric field functions to accelerate the movement of the electrons and ions in the arc column. As long as a velocity component exists vertically to the magnetic field, the charged particles in the arc gap will exhibit a circular motion in the magnetic field, and the velocity component parallel to the magnetic field will move unaffected by the magnetic field, this superposition effect is that the charged particles are moving in a spiral around the magnetic field. The Larmor radius r_{L} and rotation period T_{L} can be expressed as follows:

Fig. 4.19 Motion of the charged particles in AMF

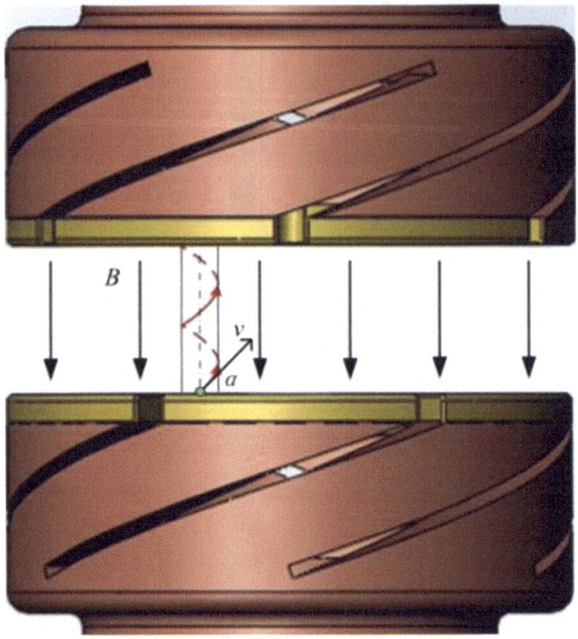

$$\begin{cases} r_L = \frac{mv_\perp}{|q|B} \\ T_L = \frac{2\pi m}{|q|B} \end{cases} \quad (4.7)$$

Among which, $v_\perp$ is the component which is vertical with respect to the AMF. The rotation direction can be determined by Lenz's law, and the winding directions of electrons and ions are the opposite. The electrons and ions are emitted from the cathode spots and move towards the anode. They will diffuse to the lower density area simultaneously with a certain radial velocity, as shown in Fig. 4.19.

For the electrons, the Larmor radius can be expressed as:

$$r_L = \frac{mv_\perp}{|q|B} = \frac{mv \cdot \cos\alpha}{|e|B} \quad (4.8)$$

Suppose the emission angle from the cathode surface is 60° and the velocity is 10^6 m/s. Then, according to the simulation results, when the current is 10 kA, the average magnetic flux density is 100 mT. The Larmor radius of the electrons is 2.85 $\times 10^{-2}$ mm and the period of rotation is 3.58 $\times 10^{-10}$ s. For Cu^+, if the emission velocity is 10^4 m/s, then the Larmor radius is 33.1 mm, and the cycle period is 4.16 $\times 10^{-5}$ s [8].

If the effect of the electric field on the charged particles is not considered, for the time being, the rising pitch in a cycle period is:

$$h = v \cdot \sin\alpha \cdot T_L \quad (4.9)$$

The pitch values of electron and Cu^+ are 0.31 mm and 360 mm, respectively. It can be referred that when the contact diameter is 41 mm and the spacing is 3 mm, the electrons are restrained by the AMF to spiral upward in a small cylindrical area. Comparatively speaking, the ions are less affected by the magnetic field. The AMF compels the electrons to rotate around the arc axis direction, which significantly lengthens the space travel of the electrons, reduces the number of electrons and ions overflowing from the arc gap and delays the occurrence of ion depletion near the anode. As a result, it becomes easier to maintain the balance of the plasma in the arc column.

Moreover, the centrifugal force generated by the rotation of the electrons and ions around the axis of the arc partially offset the magnetic pinch force (centripetal force) generated by the magnetic field of its own, thereby delaying the reduction of the arc diameter, and reducing the arc voltage and arc energy. It is very helpful to keep the arc diffusion.

4.3.2 Bennet Pinch Model

An arc column is composed of plasma with equal positive and negative charged particles. When current flows, due to the action of the magnetic field generated by the interaction between the inner and outer layers of the current, pressure is exerted toward the plasma center.

Suppose the section of the arc column is circular, as shown in Fig. 4.20, the arc radius is r_h, the axial length is L, the azimuthal magnetic field is B_a and the current

Fig. 4.20 Arc magnetic pinch model

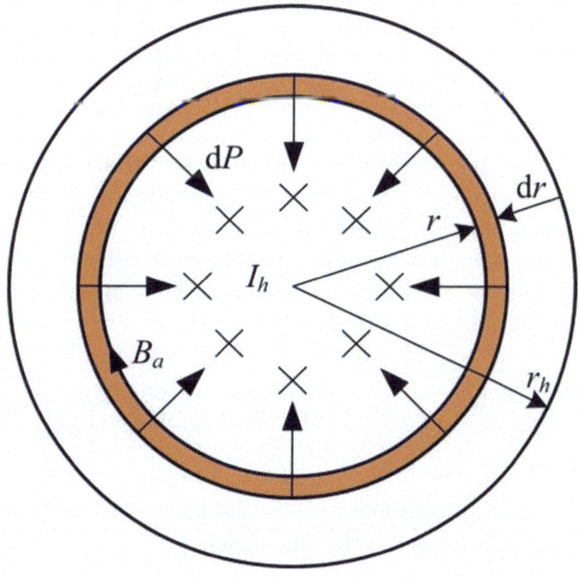

flowing in the arc column is I_h. We note that current I_h is described as the current that flows into the paper from front to back. Take a cylindrical thin layer with a radial thickness of dr at the distance r from the center of the arc column, and the current density at r is J_r, then the magnetic induction intensity conforms to the Ampere's loop theorem:

$$\oint \boldsymbol{B}_a \cdot \mathrm{d}\boldsymbol{l} = \mu_0 \sum I \tag{4.10}$$

The total current contained in the integrating loop is

$$\sum I = \int_0^r J_r 2\pi r \mathrm{d}r \tag{4.11}$$

Thus, the value of the azimuthal magnetic flux density is

$$B_a = \frac{\mu_0}{r} \int_0^r J_r r \mathrm{d}r \tag{4.12}$$

When the arc column is stationary, the Lorentz force density is balanced by the pressure gradient according to the force balance equation.

$$\nabla P + \boldsymbol{J}_r \times \boldsymbol{B}_a = 0 \tag{4.13}$$

That is,

$$\mathrm{d}P = -\mu_0 \left(\frac{J_r}{r} \int_0^r J_r \mathrm{d}r \right) \mathrm{d}r \tag{4.14}$$

The negative sign in (4.14) indicates that the pressure on the inner layer is higher than that on the outer layer. The direction of the arc pinch pressure is inward, as shown in Fig. 4.20. dP is integrated from the edge of the arc column inward and assume the pressure at the edge of the arc column as P_0, and the current density J_r to be uniform, then in the arc column, the relationship between the pressure $P(r)$ and the arc radius r is expressed as follows:

$$P(r) - P_0 = \int_0^r \mathrm{d}P = \frac{\mu_0 I_h^2}{4\pi^2 r_h^2} \left(1 - \frac{r^2}{r_h^2} \right) \tag{4.15}$$

If it is specified that $r = r_h$ and $P_0 = 0$, it can be known that maximum pressure P_{max} at the center of the arc occurs,

$$P_{\max} = \frac{\mu_0 I_h^2}{4\pi^2 r_h^2} \tag{4.16}$$

Pressure P is named as the arc magnetic pinch pressure (AMPP). It can be seen from (4.16) that when the arc current is constant, the smaller arc radius r_h is, the greater center pressure $P_{\max}$ is. During the arcing period, the magnetic pinch pressure at different sections varies, which forms a plasma pressure gradient where the plasma at the center flows to a lower pressure area. Apart from this, due to the high temperature in the arc root, the contact metal will rapidly vaporize, and a metal vapor stream perpendicular to the contact surface will also be formed. Therefore, the pressure gradient formed by the AMFF may cause jets from the cathode and anode, resulting in a variety of arc column arc manifestations.

4.3.3 Pressure Gradient of the AMPP

At a frequency of 360 Hz, with the peak current to be about 10 kA, the typical arc shape in the AMF and butt contacts is shown in Fig. 4.21a, b. We can see that the arc in the AMF contact is in a diffused state with uniform diameter. The arc in the butt contacts is clearly irregular, the arc shrinks obviously, and it is in the shape of a plasma jet column. From the perspective of AMPP, according to Eqs. (4.15) and (4.16), the AMPP in the arc column is uniform and the center pressure gradient is small. However, for the butt contacts, the radius of the arc column is the smallest in the middle position, where the central pressure of the arc column is the largest, forming a larger pressure gradient. In the simplified 2D arc model shown in Fig. 4.21c, a cylindrical coordinate system is established, and the relationship between arc radius r_h and height z is $r_h(z)$.

As shown in the arc images, $r_h(z)$ can be respectively described by the following linear and parabolic functions:

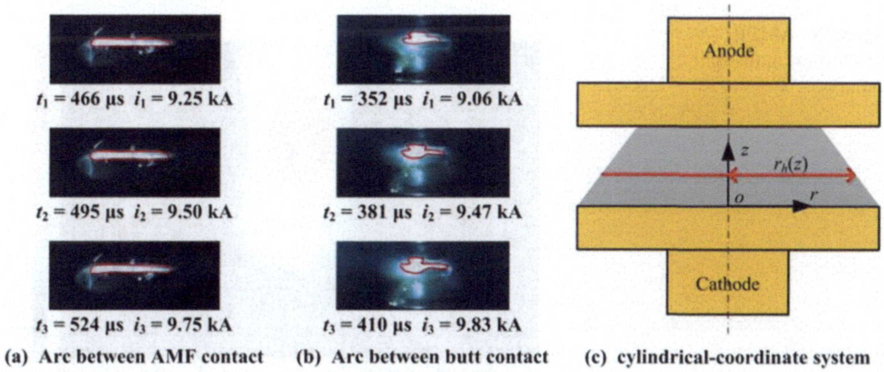

Fig. 4.21 Arc plasma appearances and simplified 2D physical model

$$r_{\text{h-AMF}}(z) = -\frac{1}{3}z + 19 \tag{4.17}$$

$$r_{\text{h-butt}}(z) = \frac{16}{9}z^2 - \frac{22}{3}z + 10 \tag{4.18}$$

When combined with Eq. (4.16), the AMPP distributions in the AMF and butt contacts are shown in Figs. 4.22 and 4.23, respectively.

For the AMF contacts, the arc radius on the anode surface is the smallest, with no obvious contraction detected. For the butt contacts, the arc is severely constricted, and the minimum arc radius is near the middle position. The maximum pressure on the anode surface of the AMF contacts can reach approximately 10^4 Pa, which is consistent with the pressure distribution for AMF contacts in [9], whereas the maximum pressure in the arc column of the butt contacts is 5×10^5 Pa. Under the pressure gradient, plasma and metal vapor flow to the lower pressure area to form a plasma flow, which strengthens the free effect of the arc gap. At the same distance, the pressure gradient along the axial direction of the butt contacts is greater, which

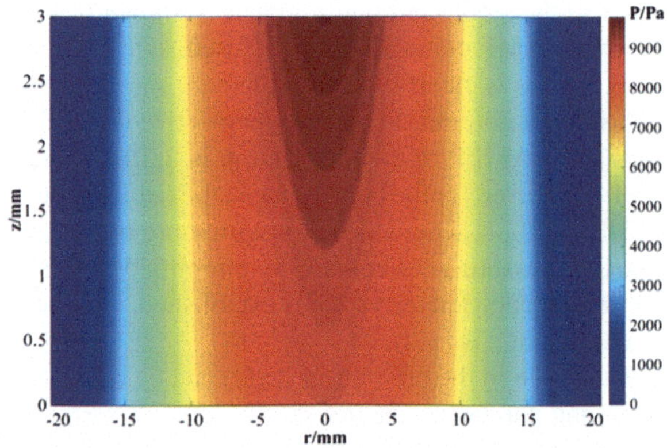

Fig. 4.22 AMPP distribution in the AMF contacts

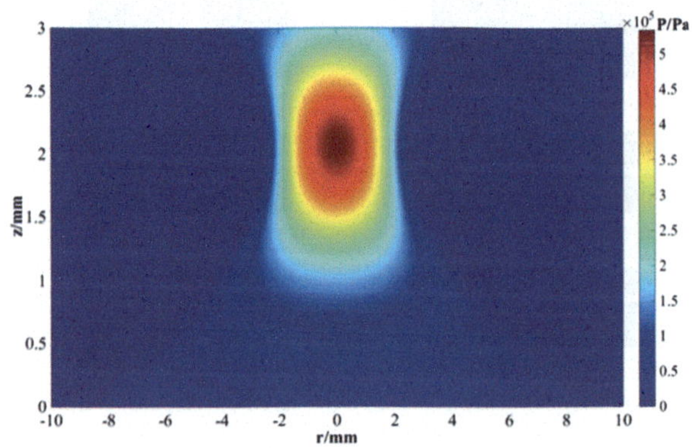

Fig. 4.23 AMPP distribution in the butt contacts

causes a large number of electrons and ions to overflow outward, and the activity in the arc area is very violent, forming a plasma jet cylindrical arc.

4.4 Reasons for the Drop-in Breaking Capacity of VF VCB

The breaking capacity for short circuit current is one of the important technical indicators of the vacuum interrupter. In order to study the breaking capacity of the vacuum interrupter at different frequencies and analyze the main factors affecting its breaking performance, VCB of AMF type with contact materials of CuCr50 and a diameter of 41 mm is selected for the experiment in this section, with an open distance of 3 mm. During the experiment, the current frequency is changed, and the current peak values of successful and failed breaking at different frequencies are obtained, as shown in Table 4.6. The $I_{success}$ column in the table indicates the maximum first-half current peak value of the successful breaking whereas the $I_{failure}$ column represents the current peak value when the breaking fails. For each frequency, the limit breaking test is carried out more than 15 times. It should be noted that when the current frequency is 500 Hz and current peak is 16.5 kA. Sometimes the breaking is successful and sometimes it isn't. The current value 16.5 kA is considered to be the critical value of breaking. There is a similar situation for 15.5 kA at 600 Hz.

As shown in Fig. 4.24, the curve × represents the current peak value at which each frequency point begins to fail, and the curve ○ represents the current peak value at which the breaking of each frequency is successful, and the breaking limit current value should be between the curve × and ○. It can be seen that as the frequency increases, the current limit breaking capacity decreases. From a macro perspective, the interruption process is determined by the competition between the RRRV and RRDS. Factors, such as di/dt at current zero (CZ), the thermal state of the contact, and the hysteresis of the AMF will affect the recovery speed of the dielectric strength. Since the RRRV du/dt, which is determined by R_0 and C_0 in Fig. 2.1, is basically the same in the experiment, this section focuses on the di/dt at CZ, the arc energy, and the hysteresis of the AMF on the current breaking.

As shown in Fig. 4.24, di/dt of CZ at different frequencies under failed interruption is represented by the curve Δ. The di/dt increases from about 50 A/μs at 360 Hz to 75 A/μs at 850 Hz. Because the di/dt has a maximum value when the current crosses

Table 4.6 Current level for successful and failed breaking at different frequency (AMF, CuCr50, 41 mm)

f (Hz)	$I_{success}$ (kA)	$I_{failure}$ (kA)
360	19.9	22.4
500	16.5	16.5
600	15.5	15.5
680	14.4	15.0
850	13.3	14.0

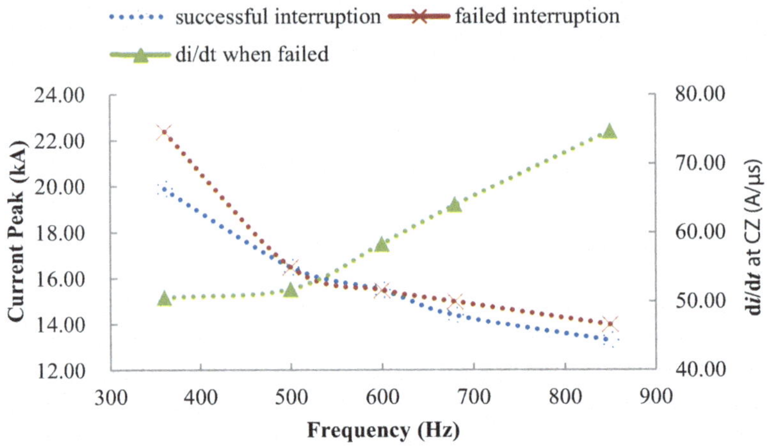

Fig. 4.24 Limit breaking capacity at different frequencies (AMF, CuCr50, 41 mm)

zero, which is related to the current peak value and frequency:

$$\left.\frac{\mathrm{d}i}{\mathrm{d}t}\right|_{\max} = \left.\frac{\mathrm{d}\left[I_{\text{peak}} \sin(2\pi f \cdot t)\right]}{\mathrm{d}t}\right|_{t=0} = 2\pi f \cdot I_{\text{peak}} \tag{4.19}$$

The above equation illustrates that when the frequency goes up, a relatively small current value will reach the same rate of change at the time of current zero-crossing. If it is considered that there is a threshold of $\mathrm{d}i/\mathrm{d}t$ in VCB, then it will be more difficult to break with the increase in frequency.

According to the Bennet pinch model, the pressure difference between the inside and the edge of the arc column is expressed in Eq. (4.15). Owing to the skin effect, with the increase of the current rate, the distribution density of current, J_r, will be low in the center and high in the surrounding area, causing the pressure at the center of the arc to increase, the shrinking effect of the arc will be more obvious. The above phenomenon can be observed in Chap. 5, which is also verified in the images of arcs that fail to start at different frequencies. The shrinking arc concentrates the arc heat and increases the energy density, which will enhance the stability of the arc. Then, when the current decreases to zero, the rate at which the amount of plasma in the cathode spots and arc gap decreases becomes slower, and there is the residual current after the arc, which will weaken the recovery strength of the medium after the arc. In summary, it can be explained that the breaking capacity at 850 Hz is less than that at 360 Hz.

According to the Mayr model, the differential equation for an arc is as follows:

$$g'(t) = \frac{\mathrm{d}g(t)}{\mathrm{d}t} = \frac{g(t)}{\tau}\left(\frac{u(t) \cdot i(t)}{P_0} - 1\right) \tag{4.20}$$

In this case, $g(t)$ is the conduction of arc, and $u(t)$ is the arc voltage, and $i(t)$ is the current. P_0 is the heat dissipation power of arc, and τ is the time constant of arc. To solve for P_0 and τ, we need to obtain the least-squares solution of Eq. (4.21),

$$\min J(\tau, P_0) = \min \sum_{j=1}^{N} \left[g'(t_j) - \left(\frac{1}{\tau P_0} g(t_j) \cdot u(t_j) \cdot i(t_j) - \frac{1}{\tau} g(t_j) \right) \right]^2 \quad (4.21)$$

According to the arc voltage and current data obtained in the experiment, the heat dissipation power and the time constant of arc under different current levels and frequencies can be calculated, as shown in Table 4.7. The experimental data for failed interruption is marked with bold and italic font. It can be seen from the data in the table that under different experimental conditions, P_0 and τ are of the same order of magnitude, indicating that the results are trustworthy. The influence of the current on P_0 is apparent. As the breaking current value increases, P_0 shows an increasing trend, as shown in Fig. 4.25a. This is because an increase in current indicates an increase in the input power of arc. Then, to ensure the power balance, the heat dissipation power will also increase. The variation of the time constant of arc is more complicated. From a macro point of view, the time constant of arc at 360 Hz is larger than that at other frequencies, indicating that the arc time has an impact on τ. At the same frequency, the time constant increases significantly in the failed interruption, as shown in Fig. 4.25b, in which successful interruption is presented by the symbol ○, and failed interruption is presented by the symbol ×. This means that the arc energy decays slowly when a failed interruption occurs, which is not conducive to breaking.

Table 4.7 Heat dissipation power and time constant of arc

f (Hz)	I (kA)	P_0 (kW)	τ (μs)
360	10.0	247.18	48.05
360	14.0	361.07	68.88
360	19.9	653.80	50.54
360	***22.4***	***875.02***	***73.35***
500	15.0	471.32	19.06
500	16.5	539.45	2.00
500	***16.5***	***549.93***	41.72
600	14.0	421.33	10.32
600	15.5	510.67	3.95
600	***15.5***	***508.90***	***38.26***
680	14.4	446.04	1.66
680	***15.0***	***516.56***	***36.02***
850	13.3	535.79	7.02
850	***14.0***	***521.02***	***40.83***

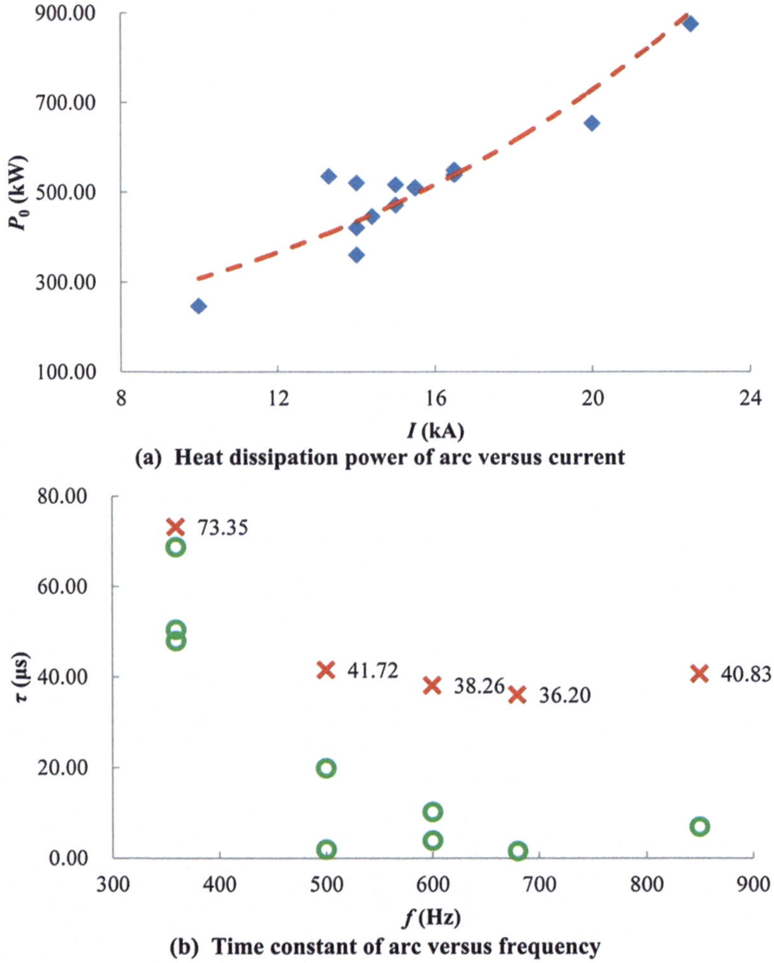

Fig. 4.25 Parameters of the Mayr model

In addition, as the current frequency rises, the eddy current effect of the AMF will be stronger, resulting in more obvious hysteresis of AMF. The AMF, B, in the contact gap is related to the current magnetic field, B_{i}, produced by the current, i, and eddy current magnetic field, B_{e}, generated by the eddy current, i_{e}, as illustrated in Fig. 2.18. According to the Ampere circuital theorem and Faraday's law of electromagnetic induction,

$$\oint_{l} \boldsymbol{B} \cdot \mathrm{d}\boldsymbol{l} = \mu_0 I \tag{4.22}$$

$$e = -\frac{d\Phi}{dt} \tag{4.23}$$

It can be drawn from Eqs. (4.22) and (4.23), the current, i, flowing through the contacts, is in the same phase with the current magnetic field, B_i. The eddy current, i_e, is induced by AMF, and the eddy current magnetic field, B_e, lags behind B by 90°. The composition of the AMF is shown in Fig. 2.18b, in which the angle, α, represents the phase of the current magnetic field, B_i, lagging behind the AMF, B.

When the current frequency is constant and the amplitude increases, B_i and B_e, will increase in proportion; thus the hysteresis phase, α, is constant. However, if the current amplitude does not change, and the frequency increases, the amplitude of B_i remains unchanged, and the amplitude of B_e increases. In order to obtain a vector synthesis relationship, the amplitude of B will decrease and the hysteresis phase α will increase, as displayed in Fig. 2.18b. Therefore, an increase in frequency will strengthen the eddy current effect and hysteresis. Moreover, the hysteresis of AMF after CZ is also a crucial element affecting reignition and breaking capacity. With regard to the vacuum arc at power frequency, Wang has simulated delay time of the AMF type contact. It is found that the maximum delay time does not exceed 0.6 ms at the contact center [9].

Converting 0.6 ms into the current phase angle, it can be found that the lagging current of AMF does not exceed 10.8° at power frequency. The ANSYS stimulation results for the AMF varying with VF current are shown in Fig. 4.26, where Q is the center point of the middle plane between the contacts and L is the diameter and the simulation current is 1 kA RMS. As the frequency is 400 Hz, the residual magnetic flux density of Q is 2.80 mT at CZ, and the hysteresis phase is about 20°. When the frequency goes up to 800 Hz, the residual magnetic flux density of Q is 3.65 mT, and the hysteresis phase even exceeds 30°. It can be seen that although the increase in frequency makes the lag time of AMF at IF smaller than the power frequency, the lag angle that truly reflects the influence of frequency on the magnetic field lag has increased tremendously. And from Fig. 4.26c, it can be seen that, before the current peak, the AMF of each point in L increases with the increase of current. However, the increase rate of the center is slow and the B is lower than that of other regions. After the peak current, the further away from the center, the faster the AMF decreases. The change process also indicates that the magnetic field of the contact center is the most affected by the eddy current, and the hysteresis of AMF is the most obvious. Therefore, residual plasmas in arc gap will be under the control of the residual magnetic field within a relatively long phase angle, and it is difficult to rapidly diffuse, which hinders the recovery of dielectric strength and affects the reignition of the arc. Hence, the hysteresis of AMF is one of the most important factors affecting the current breaking capacity for VF vacuum arc.

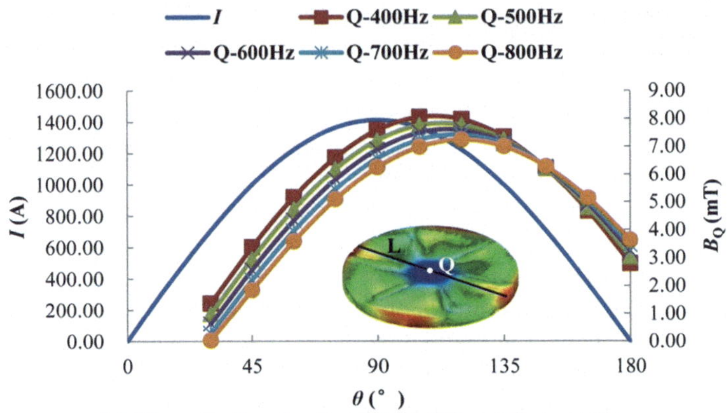

(a) Magnetic flux density of Q at different frequencies

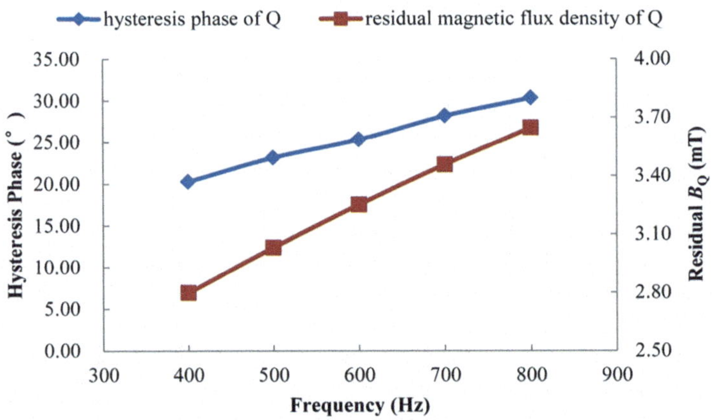

(b) Hysteresis of Q

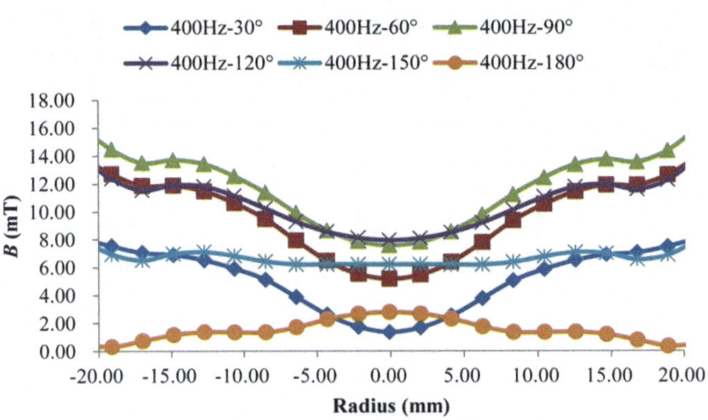

(c) Magnetic flux density of L at different phases

Fig. 4.26 Magnetic flux density of Q and L

References

1. Y. Niwa, K. Yokokura, J. Matsuzaki. Fundamental Investigation and Application of High-Speed VCB for DC Power System of Railway [C]. 24th International Symposium on Discharges and Electrical Insulation in Vacuum, Braunschweig, Germany, 2010: 125–128.
2. Yuan Jiang, Jianwen Wu, Bowen Jia. Reignition after interruption of intermediate-frequency vacuum arc in aircraft power system [J]. IEEE Access, 2018, 6: 8649–8656.
3. R. E. Voshall. Current interruption ability of vacuum switch [J]. IEEE Transactions on Power Apparatus System, 1972, PAS-91: 1219–1224.
4. Jimei Wang. Vacuum switching technology and application [M]. Beijing: China Machine Press, 2008.
5. Dingge Yang, Lijun Wang, Shenli Jia, et al. Experimental Investigation of Anode Activities in High-Current Vacuum Arcs [J]. IEEE Transactions on Plasma Science, 2010, 38(2): 206–213.
6. E. D. Taylor, P. G. Slade, M. B. Schulman. Transition to the diffuse mode for high-current drawn arcs in vacuum with an axial magnetic field [J]. IEEE Transaction on Plasma Science, 2003, 31(5): 909–917.
7. Z. Zaluki, J. Janiszewski. Transition from constricted to diffuse vacuum arc modes during high ac current interruption [J]. IEEE Transaction on Plasma Science, 1999, 27(4): 991–1000.
8. J. Kutzner, H. C. Miller. Integrated ion flux omitted from the cathode spot region of a diffuse vacuum arc [J]. Journal of Physics D: Applied Physics, 1992, 25(4): 686–693.
9. X. Huang, L. Wang, J. Deng, S. Jia, K. Qin, Z. Shi. Modeling of the anode surface deformation in high-current vacuum arcs with AMF contacts [J]. Journal of Physics D: Applied Physics, 2016, 49(7): 075202.

Chapter 5
Post-arc Breakdown in Variable Frequency Vacuum Arc

The arc voltage is a macroscopic reflection of vacuum arc plasma, which can reflect the evolution of arc and is closely related to the breaking capacity of the current. Research on vacuum arc at power frequency has shown that voltage noise would be produced at high current level. This chapter will study the phenomena of voltage spike, voltage noise, and post-arc breakdown of VF vacuum arc.

5.1 Voltage Spike and Noise

5.1.1 Arc Appearance Related to Voltage Spike and Noise

Figures 5.1 and 5.2 show the arc voltage and current of the Cu-W-WC and CuCr50 vacuum chamber with a diameter of 41 mm when the current interruption is successful and fails (current frequency is about 360 Hz).

For the Cu-W-WC contacts, when the peak current of the first half-wave is 17.5 kA, as shown in Fig. 5.1a, the current is interrupted successfully. The arc voltage is gentle, the maximum value of the voltage is 54.7 V and there is post-arc high frequency oscillation in the voltage. When the peak current of the first half-wave reaches 20 kA, as shown in Fig. 5.1b, the voltage has an obvious spike, the arc fails to break, the peak value is 83.3 V. After zero, there are reverse breakdown and significant voltage noises.

For the CuCr50 contacts, when the peak current of the first half-wave is 20 kA, as shown in Fig. 5.2a, the current is interrupted. The maximum value of the voltage is 60 V. While the peak current of the first half-wave rises to 22.5 kA, as shown in Fig. 5.2b, the arc cannot be broken. The voltage spike is 74.7 V and the reverse breakdown occurs after zero. Comparing to the Cu-W-WC contacts, the amplitudes of the voltage spike and noises are smaller and the duration is shorter in the CuCr50 contacts. When the first peak current is 22.5 kA, Fig. 5.2c shows the enlarged waveform of the recovery voltage of CuCr50 contacts.

Y. Jiang and Q. Li, *Vacuum Circuit Breaker for Aviation Variable Frequency Power System*, Power Systems, https://doi.org/10.1007/978-981-33-4781-6_5

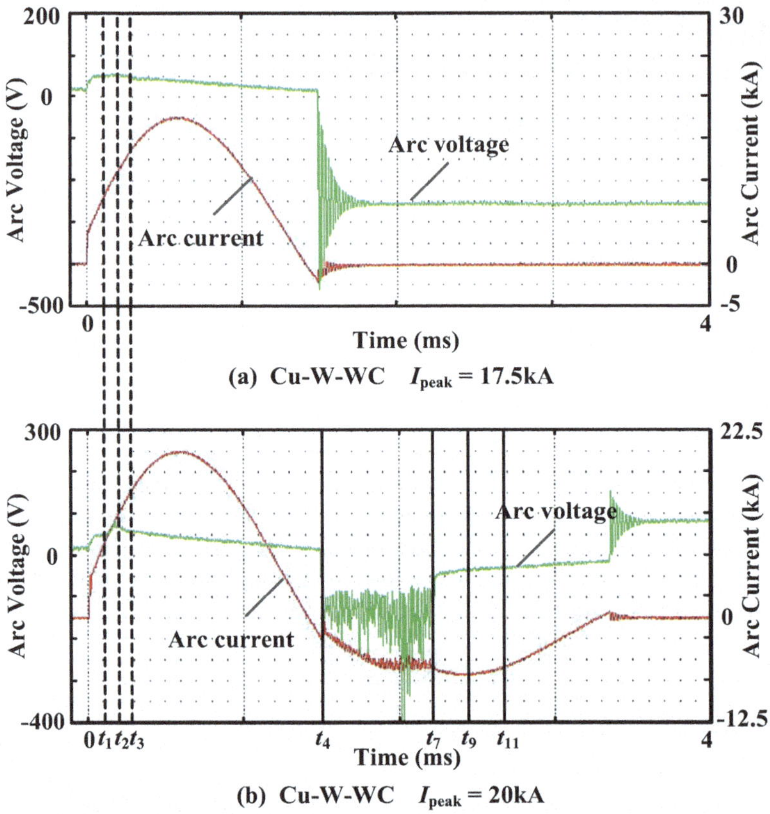

Fig. 5.1 Arc voltage and current of Cu-W-WC contact at 360 Hz

From the comparison of the above several sets of data, it can be concluded as follows: the Cu-W-WC and CuCr50 contacts with the same diameter, have the same current and voltage curves when current is successfully interrupted, and the experimental phenomena when they are not broken are basically the same, but the arc voltage spike and noises of the CuCr50 are lower. The interruption capacity in CuCr50 is stronger than that in Cu-W-WC. With the same peak current of the first half-wave, peak voltage of the CuCr50 contacts is lower. This is because, compared to W and WC, on one hand, Cr has a lower boiling point and a higher saturated vapor pressure, and it can generate enough metal vapor at a relatively low temperature; on the other hand, the thermal conductivity is low, and the heat is not easy to conduct away. The temperature is easy to rise. The two factors work together to produce a sufficient amount of metal vapor to maintain the arc with a lower energy input, so the arc voltage peak and noise between the CuCr50 contacts during arcing are relatively low.

Arc images at the moment of the first current peak are displayed in Fig. 5.3. For the small-open-distance VF vacuum arc, no concentrated state is observed, and the

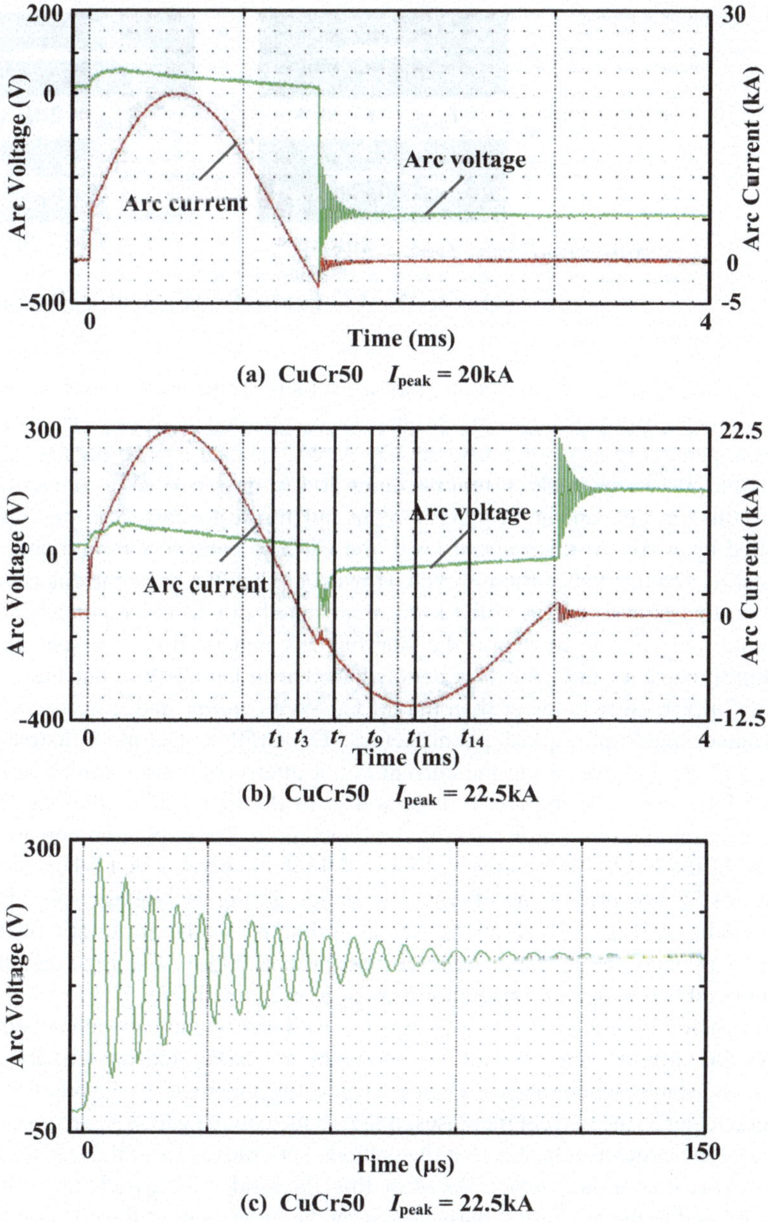

Fig. 5.2 Arc voltage and current of CuCr50 contact at 360 Hz

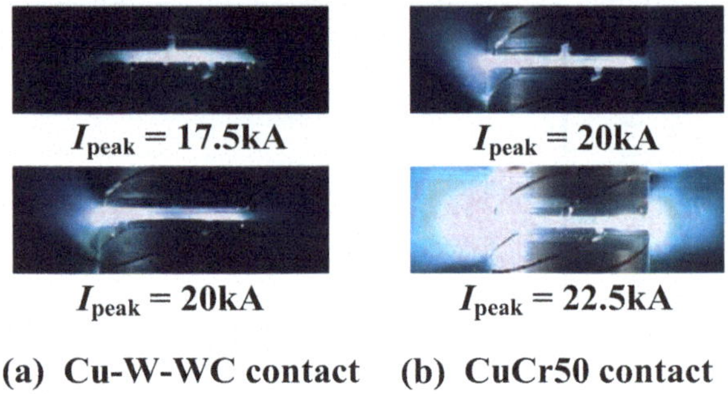

Fig. 5.3 Arc images at peak current

arc is in a diffuse state. For the same kind of contacts, as the current level increases, the arc diameter becomes larger and brighter, accompanied by obvious arc overflow. When the peak current is 17.5 kA, the Cu-W-WC contact can be opened and the arc is stable. When the peak current becomes 20 kA, the Cu-W-WC contact fails to open. At this time, plasma in the gap is violent and the arc appears to be incompletely controlled by AMF. At this current level, the CuCr50 contacts can be opened, the arc is stable, and the distribution is even. However, when the peak current continues to rise up to 22.5 kA, arc is failed to be interrupted in CuCr50 contacts. The arc is extremely bright and moves to the side of the contacts. It can be seen that the interruption capacity in CuCr50 is better than that in Cu-W-WC, but the melting point of CuCr50 alloy is lower than that of Cu-W-WC alloy, and it is easy to melt, which makes the droplet splash phenomenon of CuCr50 contact more distinct.

As mentioned above, when the current is not interrupted successfully as shown in Figs. 5.1b and 5.2b, the arc voltage spikes in the first half of the wave; after the arc is reignited, noise appears in the voltage curve. These phenomena are more apparent in the Cu-W-WC contacts. Figure 5.4a demonstrates that when the peak current is 17.5 kA, there is no obvious arc image during the voltage spike period. Correspondingly, Fig. 5.4b shows at the same time point, the arc image of the arc voltage peak when the first half-wave current peak is 20 kA. The time and current value marked below the arc image correspond to the time in Fig. 5.1.

When there is no noticeable arc spike, arc is well distributed between the contacts gap. As the current increases, the arc diameter increases, and the distribution is uniform and stable; when there is a voltage spike, the arc moves to the outer space of the contact, and as the current increases, it leaves the contact gap. The arc burning on the side of the contact indicates that the cathode spot moves from the surface of the contact to the side of the contact. Based on this, the sudden increase in arc voltage is closely related to the arc partial burn caused by the movement of the cathode spot to the side of the contact [1]. In Fig. 5.4b, the voltage spike appears around the moment of t_2. At this time, the overflowing arc rotates and is in an irregular shape. As the breaking current value increases, the number of plasma and metal vapor between the arc gaps increase. In AMF, in the case of high current, the arc column area receives a strong Lorentz force and the spreading arc rotates clockwise.

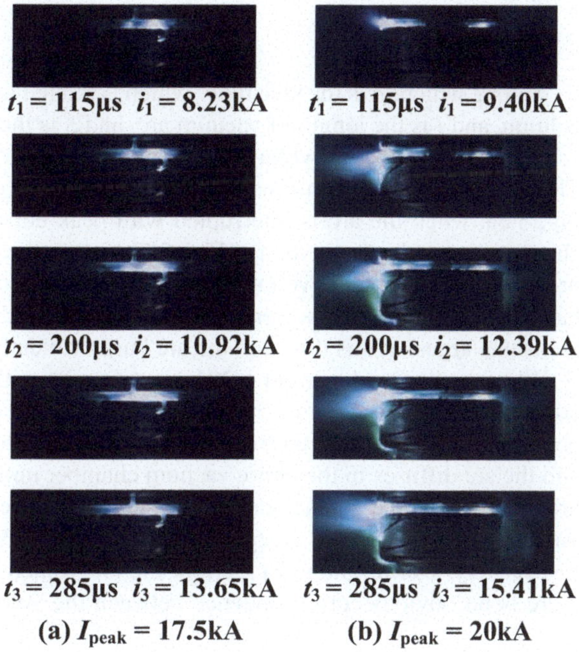

Fig. 5.4 Arc images between Cu-W-WC contacts with and without spikes

5.1.2 Reasons for the Voltage Spike and Noise

The movement of the arc outside the arc gap may be an important factor causing the voltage spike. Arc voltage U_{h} is determined by voltage drops in electrode fall and arc column, as shown in Eq. (5.1),

$$U_{\mathrm{h}} = U_0 + U_{\mathrm{Z}} \tag{5.1}$$

Among which, U_0 is the voltage drops in electrode fall, which is related to contact material. And U_{Z} is the voltage drops in arc column, which is related to the plasma density in arc column and arc resistance.

The arc column voltage drop U_{Z} is related to current and arc resistance R,

$$U_{\mathrm{Z}} = I \cdot R \tag{5.2}$$

Under normal conditions, vacuum arc voltage in short gap is mainly determined by the voltage drops in electrode fall, and the voltage drops in arc column approach zero. By comparison, it can be observed in Fig. 5.4 that when the arc burns on the side of the contact, the arc path is bent, which is equivalent to an increase in the length of the arc column, while the arc shrinks and the cross-sectional area becomes smaller. The resistance equation is (5.3):

$$R = \rho \cdot \frac{l}{S} \tag{5.3}$$

Among which, ρ is the resistivity which is related to the plasma density in the arc column, and l is the length of vacuum arc, and S is the area of vacuum arc.

As is known, the arc voltage increases due to the increase of arc resistance R. This inference can be confirmed from the waveforms of Rover time in Fig. 5.5: For Fig. 5.5a, when the arc is interrupted with peak current 10 kA, the arc resistance changes smoothly; however for Fig. 5.5b, when the peak current 20 kA, the arc breaking fails. When the voltage spikes, it can be observed that the arc resistance R has a spike, and the arc power P also has a spike, indicating that the energy of the arc is the largest at this time and the arc burns violently.

When the first peak current is 20 kA, are images with voltage noises in Cu-W-WC contacts are displayed in Fig. 5.6. The noise frequency is about 100 kHz, and lasts for about 700 μs. It can be seen from the arc image that the voltage noise starts from t_4, and the arc diffuses in the entire vacuum chamber until the noise ends at t_7, and the arc reappears in the contact gap. From the perspective of voltage value, the voltage values, at the moment of t_4, t_5 and t_6, are −280.67 V, −205.33 V, and −85.33 V, respectively. During the noise period, the arc voltage fluctuates high and low, and there is no obvious correspondence between the voltage value and the arc image.

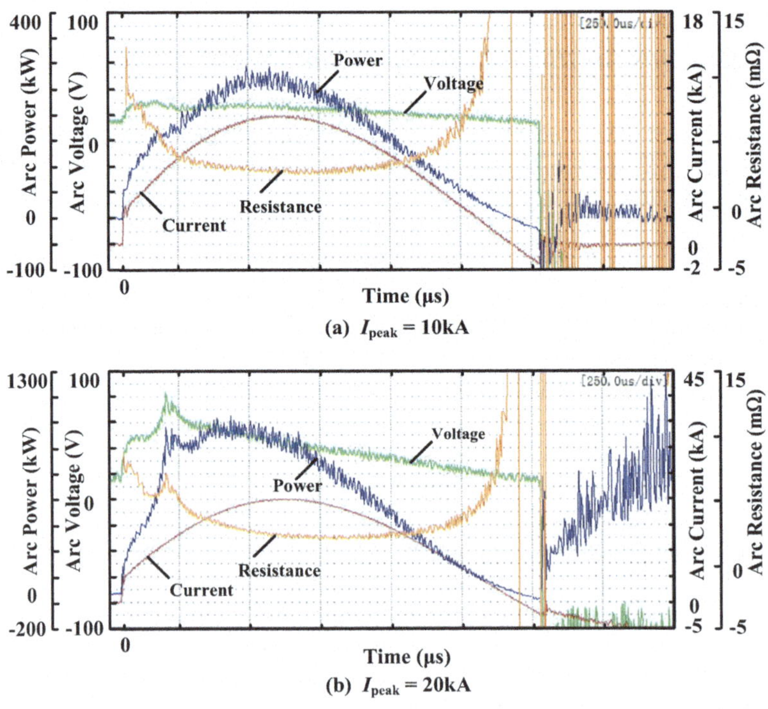

Fig. 5.5 Waveforms of power and resistance

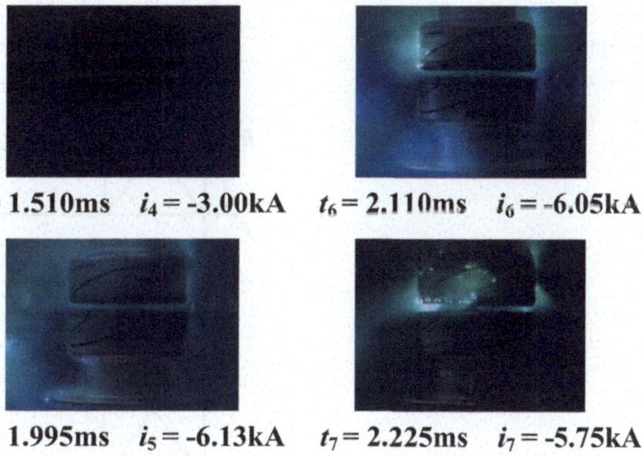

Fig. 5.6 Arc images in the periods of arc voltage noise between Cu-W-WC contacts

When the noise ends at t_7, the arc voltage is −62.67 V. After the high current, the residual plasma density between the arc gaps is large, making the arc run out of the arc gap and diffuse in the arc chamber, which may be the direct cause of the noise. Due to the addition of W and WC, the temperature of Cu-W-WC contact becomes very high. Therefore, the ability to emit thermionic electrons after the arc is very strong and the arc is easy to break down. The noise is more obvious than CuCr50 contact.

Besides, in Fig. 5.5b, it can be observed that the VF vacuum arc not only moves outside the arc gap but also the cathode emission area appears at the root of the contact. This is different from the case that the anode energy is concentrated when the current is large at power frequency and the arc is likely to concentrate in the center of the anode contact [2]. Compared with the power frequency current, the increase in frequency causes the voltage drop on the inductance of the contact excitation coil to increase. The contact coils of anode and cathode in this experiment are equivalent to inductance, and the following estimations are made:

$$U_{\mathrm{L}} = \omega L I = 2\pi f \cdot B \cdot S = 2\pi f \cdot B \cdot \pi R^2 \tag{5.4}$$

Among which, f is the frequency of the current. R is the radius of the cup-type contact. B is taken as the magnetic induction at the axis center of the contact for a simplified calculation.

In the contact structure shown in Fig. 5.7, the AMF generated by the contact coil can be equivalent to the algebraic sum of B_1 generated by the forward coil with a length of $2L$ and B_2 generated by the reverse coil with the length of $2l$,

$$B = B_1 + B_2 \tag{5.5}$$

B_1 and B_2 of any point z on the axis can be obtained according to the following equations,

Fig. 5.7 Equivalent structure diagram of AMF contacts

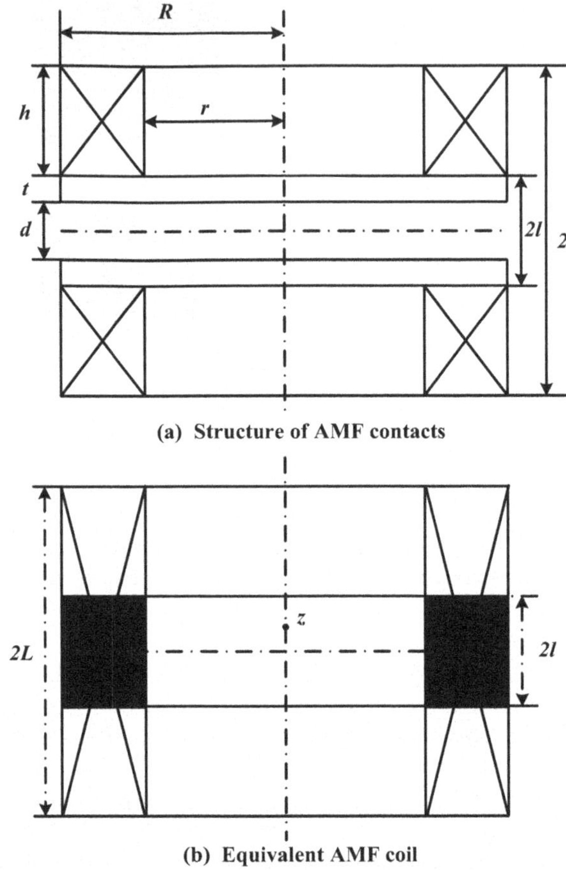

$$B_1 = \frac{1}{2}\mu_0 J\left[(L-z)\cdot\ln\frac{R+\sqrt{R^2+(L-z)^2}}{r+\sqrt{r^2+(L-z)^2}} + (L+z)\cdot\ln\frac{R+\sqrt{R^2+(L+z)^2}}{r+\sqrt{r^2+(L+z)^2}}\right] \tag{5.6}$$

$$B_2 = -\frac{1}{2}\mu_0 J\cdot\left[(l-z)\cdot\ln\frac{R+\sqrt{R^2+(l-z)^2}}{r+\sqrt{r^2+(l-z)^2}} + (l+z)\cdot\ln\frac{R+\sqrt{R^2+(l+z)^2}}{r+\sqrt{r^2+(l+z)^2}}\right] \tag{5.7}$$

$$\begin{cases} L = \frac{d}{2} + h + t \\ l = \frac{d}{2} + t \\ J = \frac{\sqrt{2}NI}{(R-r)h} \end{cases} \tag{5.8}$$

Among which, d is the open distance. h is the height of contact coil. t is the thickness of contact blade. R is the outer radius of the cup-type contact and r is the inner radius. N is the number of coil turns. I is the current and J is the current density.

When z is taken in the middle plane of the open distance, i.e., $z = 0$, the relationship is as follows,

$$\begin{cases} B_1 = \mu_0 J \cdot L \cdot \ln \dfrac{R + \sqrt{R^2 + L^2}}{r + \sqrt{r^2 + L^2}} \\ B_2 = -\mu_0 J \cdot l \cdot \ln \dfrac{R + \sqrt{R^2 + l^2}}{r + \sqrt{r^2 + l^2}} \end{cases} \tag{5.9}$$

Then the synthesized total magnetic field B is,

$$B = B_1 + B_2 = \mu_0 J \left[L \cdot \ln \frac{R + \sqrt{R^2 + L^2}}{r + \sqrt{r^2 + L^2}} - l \cdot \ln \frac{R + \sqrt{R^2 + l^2}}{r + \sqrt{r^2 + l^2}} \right] \tag{5.10}$$

The values of the above parameters are listed in Table 5.1.

Based on this calculation, the peak value of the contact inductance voltage is about 1.3 V. Close the Cu-W-WC contact and inject IF current with a peak value of 20 kA and a frequency of 360 Hz. The measured peak voltage value is 2 V, as shown in Fig. 5.8.

As shown in Fig. 5.1b, when the current peak is 20 kA, the maximum arc voltage is 83.3 V in Cu-W-WC contact and the average value of the arc voltage can be estimated

Table 5.1 The structural parameters of the cup-type AMF contact

Parameter	h (mm)	t (mm)	R (mm)	r (mm)	N
Value	12	2	20.5	17	1/4

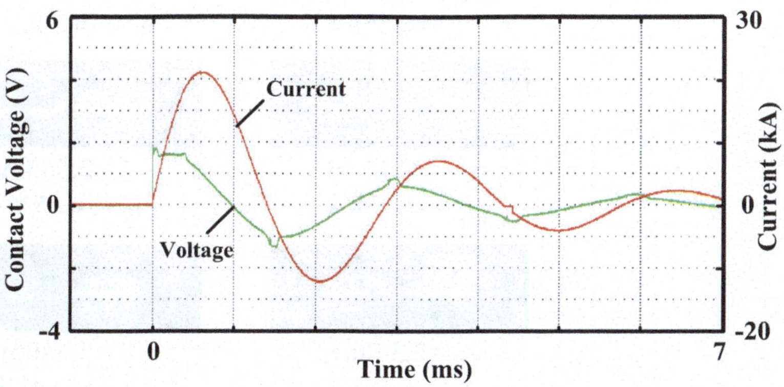

Fig. 5.8 The measured inductance voltage of a longitudinal magnetic contact

from the voltage curve to be about 40 V. Thus, it can be concluded the voltage drop value of the contact inductance can be compared with the arc voltage value. Since the metal vapor and plasma are continuously dissipated to the vacuum chamber during arcing, there are potential conductive channels and breakdown may occur at high density. Therefore, when the voltage drop on the contact coil is equivalent to the arc voltage somewhere in the space, the current can not only flow through the contact coil, but also can bypass the coil and conduct directly through the arc area outside the contact, which makes the IF vacuum arc not only overflows the arc gap, but the area emitted by the cathode appears at the root of the contact.

5.2 Post-arc Breakdown of VF Vacuum Arc

When the VF current fails to be broken, the arc experiences a process from extinguishing to post-arc breakdown. Figure 5.9 shows the post-arc breakdown process of the VF vacuum arc between the AMF contacts of Cu-W-WC with a diameter of 41 mm. The current frequency of the first half-wave is 360 Hz, and the peak current is 20 kA. The time and current values in Fig. 5.9 correspond to the time marked in Fig. 5.1. Time t_6 to t_{11} is the second half-wave after breakdown, and the current is reversed. The electrode at the top in Fig. 5.9 is the new cathode. It can be observed that the breakdown point of the arc appears at the edge of the contact. During the breakdown process, the arc expands and has a tendency to gradually shrink towards the anode. There are many cathode spots, which not only appear on the surface of the contact but also spread to the outside of the contact.

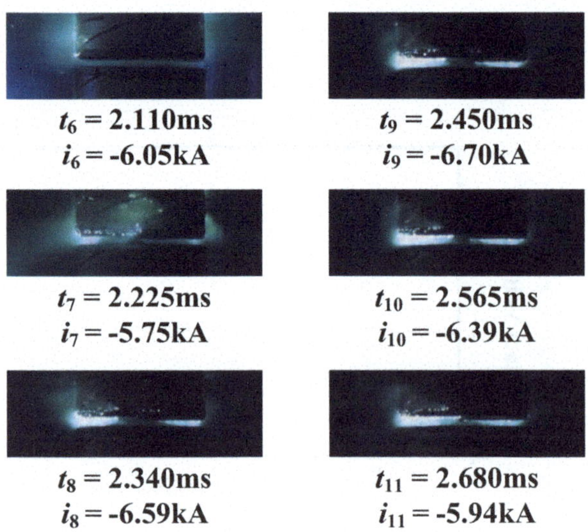

Fig. 5.9 Process of IF arc breakdown in Cu-W-WC contacts (20 kA, 360 Hz)

Figure 5.10 shows the post-arc breakdown process of the IF vacuum arc between the AMF contacts of CuCr50 with a diameter of 41 mm. The current frequency of the first half-wave is 360 Hz, and the peak current is 22.5 kA.

During the breakdown, with the increase of the current, the arc gradually expands, and the breakdown point of the arc first appears at the edge of the contact, which is from the breakdown process of Cu-W-WC contact and the obvious droplet splashing can be seen from the arc image. Time t_1 to t_3 is the first half-wave of the current, and the upper electrode in Fig. 5.10 is the anode. Time t_4 to t_{14} is the second half-wave after the arc is reignited, and the upper electrode is the new cathode. It can be observed that the number of metal droplets ejected from the cathode contact is more than that ejected from the anode contact. After the first half arc is extinguished, the temperature in the chamber is high and the density of the remaining metal vapor and plasma is high. After reignition, new energy is injected, resulting in more and larger metal droplets.

In the case of high current, the temperature of arc is high. Density of plasma in space is high and the plasma diffuses slowly, therefore, the conductivity between the contacts becomes very strong, it is very good at electric conduction between contacts. For the rate of dielectric recovery process is slower than the voltage rise

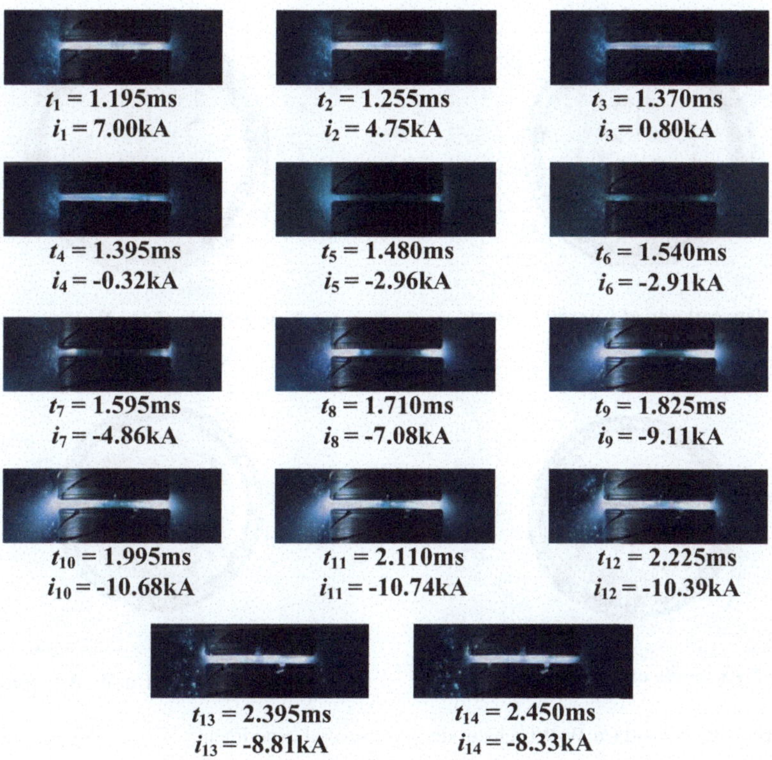

Fig. 5.10 Process of IF arc breakdown in CuCr50 contacts (22.5 kA, 360 Hz)

process, which causes the arc to re-burn. Although the diffusion rate of post-arc plasma is fast [3]: it is nearly 1–2 × 10^4 m/s for metal atomic and molecular and 10^6 m/s for electron, making the diffusion time in the level of microseconds, the temperature of the metal contact surface after arc can reach several thousand degrees and the cooling rate cannot be as fast as microsecond level. Electrons are easy to be emitted from an old anode with high temperature, thus the old anode turns to be a new cathode. The high-temperature original anode easily emits electrons and becomes a new cathode. The evaporation is high in the contact surface, which leads to break down after the arc extinction [4].

The melting point of W and WC is around 3000 °C. For Cu-W-WC contacts, due to the high arc temperature and high energy, their capability to emit thermionic electrons after arc is extremely strong. It will result in a high residual plasma density in the contact gap after current zero-crossing, making the arc easier to reignite. For CuCr50 contacts, since the melting point of Cr is 1875 °C, the ablation resistance is not as good as Cu-W-WC contacts, which can be clearly noticed from the comparison in Fig. 5.11. But CuCr50 has low temperature and high vapor pressure, and its post-arc energy is relatively weak. Therefore, even if metal droplets are ejected after the arc, the electrical conductivity of the space in the chamber is still not as strong as

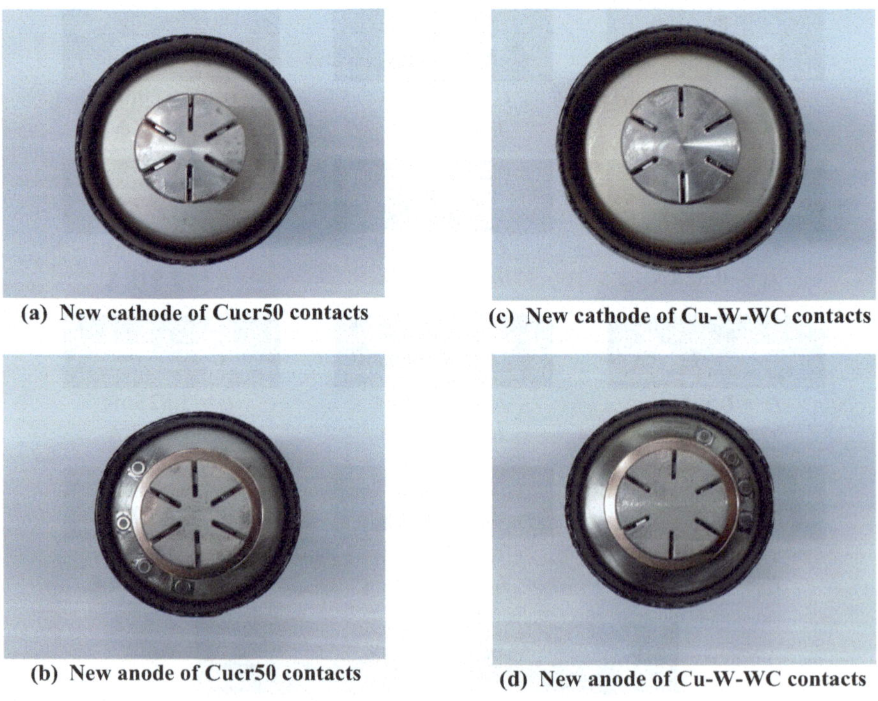

(a) New cathode of Cucr50 contacts

(c) New cathode of Cu-W-WC contacts

(b) New anode of Cucr50 contacts

(d) New anode of Cu-W-WC contacts

Fig. 5.11 CuCr50 and Cu-W-WC electrodes after several experiments that of Cu-W-WC. This is the reason why its breaking capacity is stronger than that of Cu-W-WC alloy.

5.3 Analysis of Post-arc Breakdown

5.3.1 *Position of Post-arc Breakdown*

When interruption failed at 360 Hz and 22.4 kA, the post-arc breakdown process in CuCr50 contacts is shown in Fig. 5.12a. Before current zero, the upper contact is anode and the lower contact is cathode in the arc images from t_1 to t_3. Since t_4, the current decreases to zero and is inversed, then the old anode turns to be the new cathode and the old cathode turns to be the new anode. At t_6, it can be observed

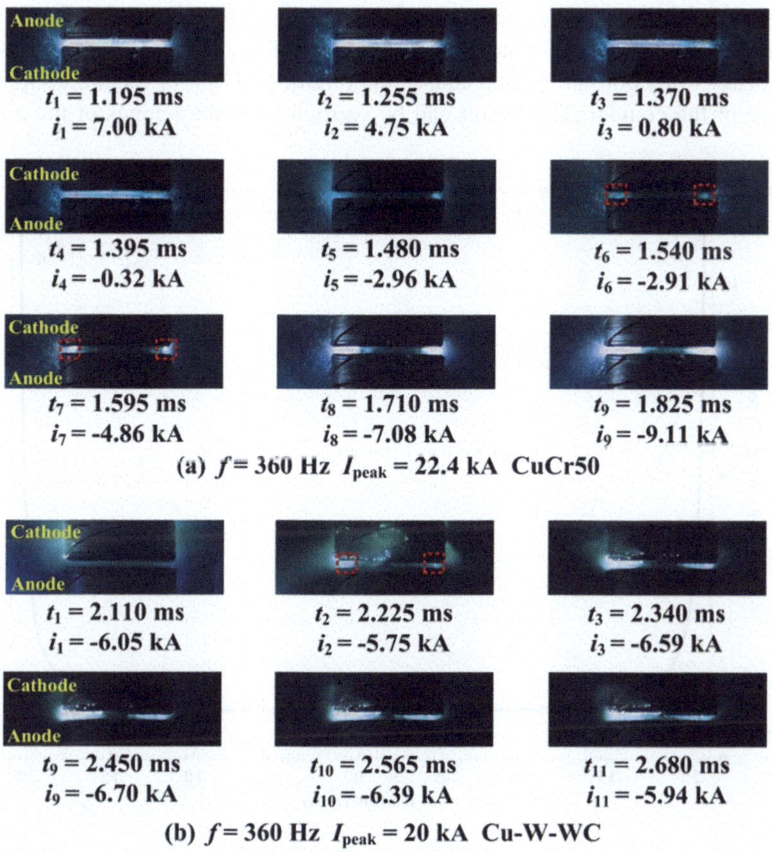

Fig. 5.12 Post-arc breakdown process in various contacts

that the position of breakdown is at the margin of the contacts. With the increase in current, the arc diameters expand and macroscopic particles sputter from the contacts in the process. In the contact with various materials, such as Cu-W-WC, it can also be observed that breakdown appeared on the contact margin first, as shown in Fig. 5.12b. Unlike the process CuCr50 contacts, there are no obvious particles at macroscopic level in the arc column. And there are not so many macroscopic particles; nevertheless, some cathode spots appear around outside the new cathode.

If the cathode and anode contacts are regarded as parallel charged plates, the Matlab program is used to simulate the charge density distribution on the surface. The result is displayed in Fig. 5.13, among which Q_0 is the datum line and Q_1 is curve of the relative charge density. It can be seen that the charge density is high due to the edge effect [4]. Given the previous experimental results, the maximum recovery voltage is about 500 V after CZ when interruption fails. Then we apply a voltage difference to the two poles of the contacts with a distance of 3 mm to simulate the electric field distribution between the contacts. The result is shown in Fig. 5.14b, in which the electric field intensity is represented by color. Due to the edge effect, the electric field intensity is concentrated at the edge of the contact. As a result, the probability of field emission increases, and the particles in this area move fast and are not easy to recombine, which causes the reignition point of the arc to appear at the edge of the contact. This result can be verified from the images of the ablative

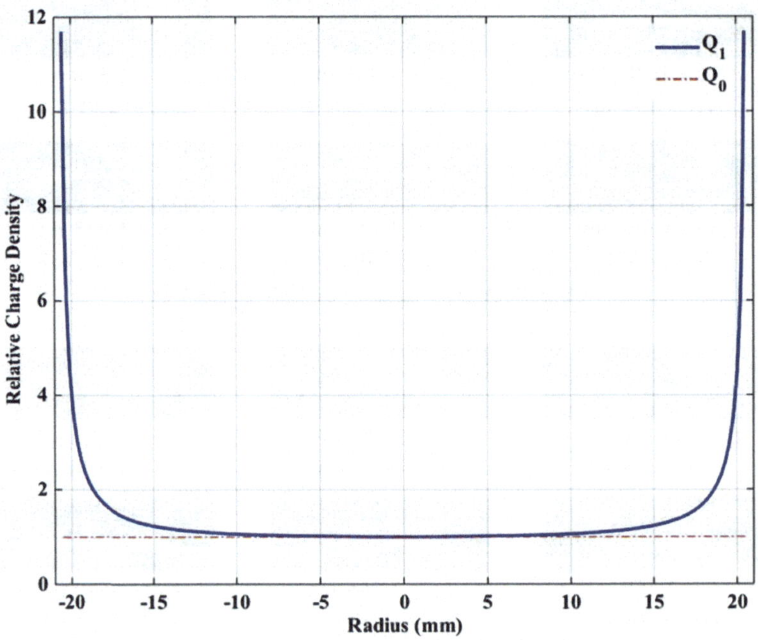

Fig. 5.13 Distribution of the charge density in the electrode surface

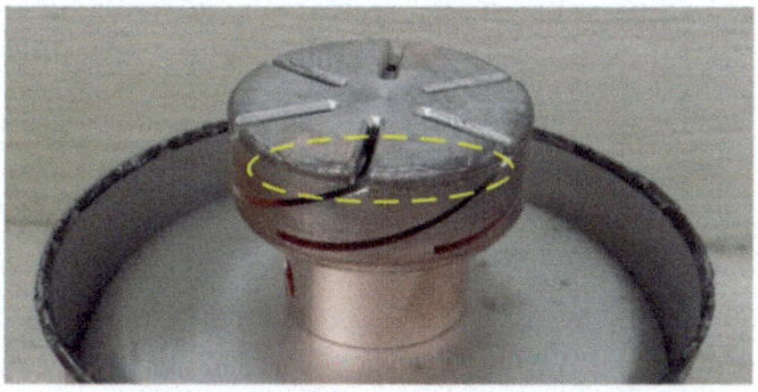

(a) Top view of the CuCr50 contacts

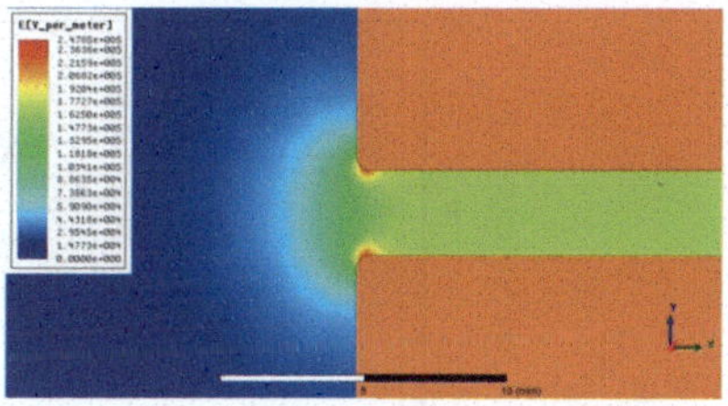

(b) Distribution of electric field in the contact distance

Fig. 5.14 Ablation and electric field at the margin of the contact

electrodes in Fig. 5.14a. After several experiments, the more severely ablated parts appear at the edge for both CuCr50 contacts.

5.3.2 Macroscopic Particles

Figure 5.15 shows the arc voltage and current waveforms when the CuCr50 contact fails to break at the current frequency of 360, 500, 600, and 680 Hz. The sputtering of macroscopic particles in CuCr50 contacts during breakdown is captured by high-speed camera, as displayed in Fig. 5.16. The time and current marked below each arc image in Fig. 5.16 correspond to the time and the current in Fig. 5.15. After reignition, the upper electrode is a new cathode, and the lower electrode is a new anode. It can be seen that there are many particles in the chamber and the sputtering is severe. At each frequency, 10 arc images were continuously recorded and the observational particle was marked with a red cycle.

Taking the contact diameter of 41 mm as a reference, the coordinate values of droplets captured at various times under different frequencies are recorded in Tables 5.2, 5.3, 5.4 and 5.5. The distance to the origin s is calculated according to Eq. (5.11):

$$s = \sqrt{x^2 + y^2} \tag{5.11}$$

According to the *xoy* coordinate system, the movement trajectory of each droplet is drawn as shown in Fig. 5.17, in which the arrow indicates the direction of movement.

Based on the moving trace in two dimensions, the motion process is not greatly affected by the frequency. After being ejected from the contact, the droplet moves in an approximately straight line β (referring to the horizontal direction) towards the shield. The displacement–time curves for each droplet are shown in Fig. 5.18a. It can be found that there is an acceleration process, A, and a uniform speed process, B, as indicated by the dotted line in every curve. The acceleration process A is expressed in Fig. 5.18b and the uniform motion process B in Fig. 5.18c, and the equations for

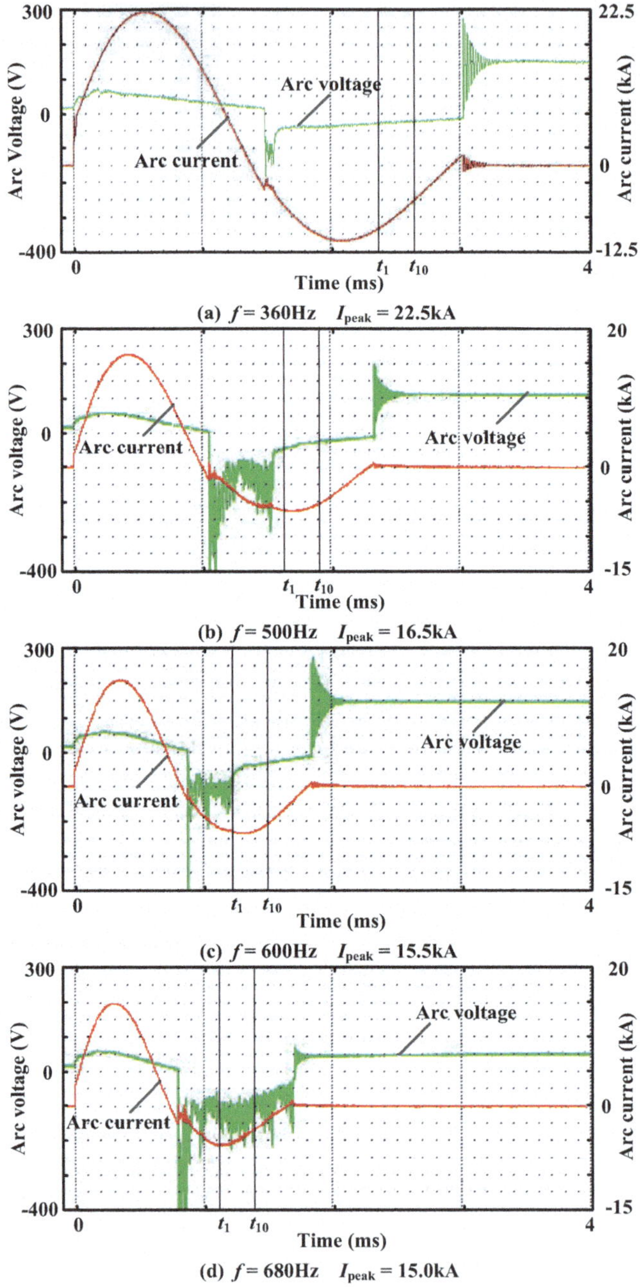

Fig. 5.15 Arc voltage and current in CuCr50 contacts during failed interruption

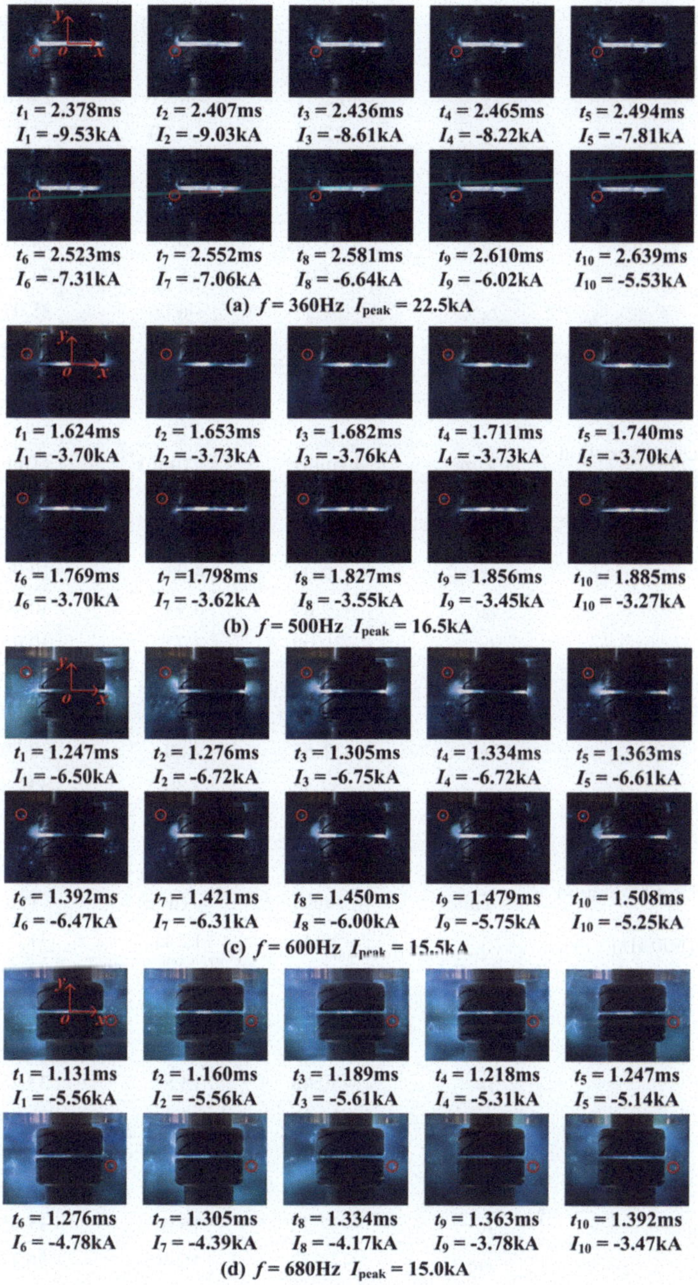

Fig. 5.16 Macroscopic particles in CuCr50 contacts during breakdown

Table 5.2 Droplet motion coordinates during breakdown (360 Hz)

Time	x (mm)	y (mm)	s (mm)
t_1	−22.32	−4.62	22.79
t_2	−22.57	−4.91	23.10
t_3	−22.82	−5.20	23.41
t_4	−23.32	−5.20	23.89
t_5	−23.57	−5.49	24.20
t_6	−23.82	−5.77	24.51
t_7	−24.08	−5.77	24.76
t_8	−24.33	−6.06	25.07
t_9	−24.58	−6.35	25.38
t_{10}	−24.83	−6.35	25.63

Table 5.3 Droplet motion coordinates during breakdown (500 Hz)

Time	x (mm)	y (mm)	s (mm)
t_1	−28.09	6.35	28.80
t_2	−28.34	6.64	29.11
t_3	−29.09	6.93	29.90
t_4	−29.60	6.93	30.40
t_5	−30.09	7.22	30.95
t_6	−30.60	7.51	31.50
t_7	−31.10	7.51	31.99
t_8	−31.85	7.80	32.79
t_9	−32.35	7.80	33.28
t_{10}	−32.85	7.80	33.77

Table 5.4 Droplet motion coordinates during breakdown (600 Hz)

Time	x (mm)	y (mm)	s (mm)
t_1	−28.46	12.41	31.05
t_2	−28.97	12.70	31.63
t_3	−29.22	12.99	31.97
t_4	−29.97	13.28	32.78
t_5	−30.48	13.86	33.48
t_6	−30.97	13.86	33.93
t_7	−31.22	14.15	34.28
t_8	−31.72	14.72	34.97
t_9	−32.23	15.01	35.55
t_{10}	−32.73	15.30	36.13

Table 5.5 Droplet motion coordinates during breakdown (680 Hz)

Time	x (mm)	y (mm)	s (mm)
t_1	22.70	−4.04	23.05
t_2	23.20	−4.04	23.55
t_3	23.70	−4.33	24.09
t_4	24.20	−4.62	24.64
t_5	24.70	−4.62	25.13
t_6	25.20	−4.91	25.68
t_7	25.45	−5.20	25.98
t_8	25.71	−5.20	26.23
t_9	26.21	−5.49	26.77
t_{10}	26.46	−5.77	27.08

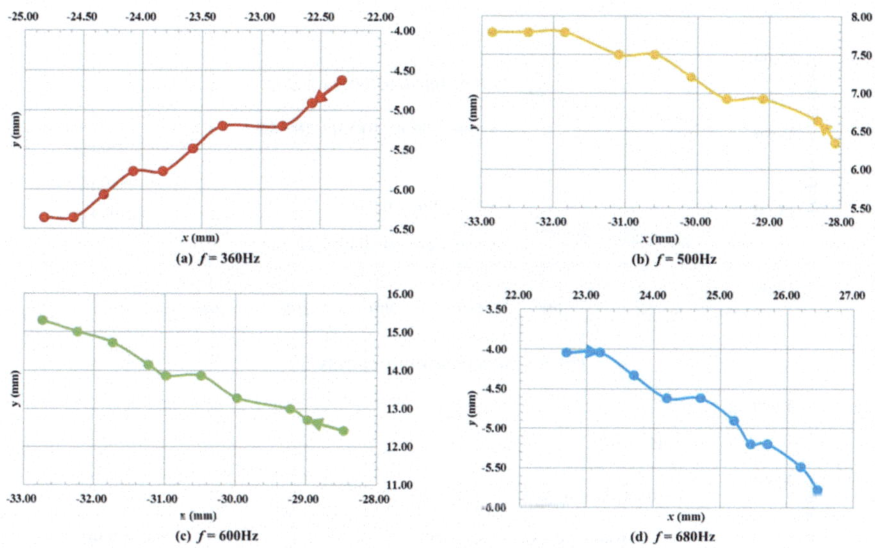

Fig. 5.17 Moving trace of droplets at different frequencies

displacement–time of the movement are obtained using quadratic fitting and linear fitting methods.

The above process can be explained as follows: there are electrons, ions, metal vapor, and droplets in the arc area, so the pressure in the center is greater than that near the shield, which can drive the droplets to accelerate [5]. In the former stage, the droplets have just left the contacts and the arc gap where the pressure is relatively high, so the droplet accelerates. In the latter stage, the droplets have already left the gap and the effect of pressure on the movement is weakened. Thus, it can be regarded as a process of uniform motion. Based on the data, both the acceleration

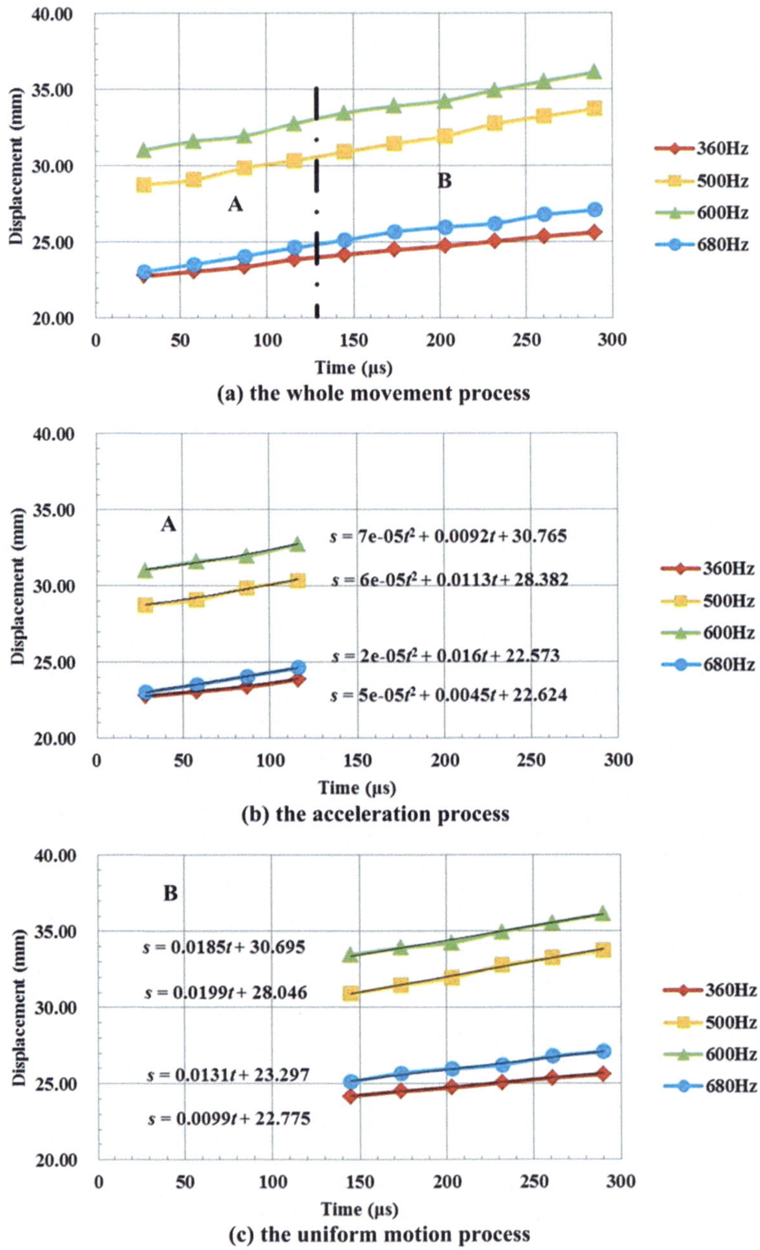

Fig. 5.18 Curves of displacement–time for droplets at different frequencies

Table 5.6 Parameters of droplet movement

f (Hz)	β (°)	v_{avg} (m/s)	a (m/s^2)
360	13.24	10.86	10×10^4
500	13.40	19.04	12×10^4
600	24.30	19.44	14×10^4
680	10.99	15,43	4×10^4

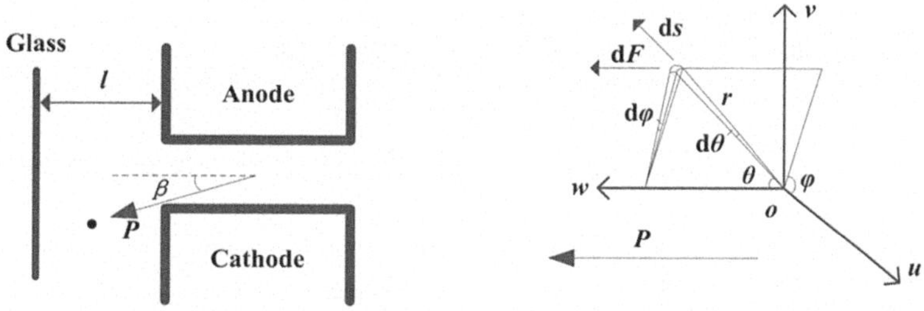

(a) Schematic diagram for droplet and pressure in vacuum chamber

(b) Spherical coordinate system inside a droplet

Fig. 5.19 Movement and pressure of metal droplet in vacuum chamber

in the accelerating process and the average velocity v_{avg} in the entire process can be approximated. The specific values are listed in Table 5.6.

According to the calculation results, the metal droplet has an acceleration of 10^4 m/s^2 during its movement, which is bound to be affected by external forces. If the ejected metal droplet is regarded as spherules, the pressure P is set according to the direction β of the droplet movement, as shown in Fig. 5.19a.

Suppose the radius of the metal droplet is r, establish a spherical coordinate system inside it, and set the pressure direction as $+w$. As shown in Fig. 5.19, the force dF on the ds is:

$$\mathrm{d}F = P \cdot \mathrm{d}s \cdot \cos\theta \tag{5.12}$$

The cell area ds in the spherule coordinate system can be expressed as:

$$\mathrm{d}s = r \cdot \sin\theta \cdot \mathrm{d}\varphi \cdot r \cdot \mathrm{d}\theta \tag{5.13}$$

If the particulate is small enough, the pressure gradient in the w direction can be considered unchanged within $[-r, r]$. Because only the external force generated by the pressure difference is considered, the pressure is P_0 where $w = 0$, and the pressure P at any angle is:

$$P = P_0 + \frac{\partial P}{\partial w} \cdot w = P_0 + \frac{\partial P}{\partial w} \cdot r \cdot \cos\theta \tag{5.14}$$

By substituting Eqs. (5.13) and (5.14) into Eq. (5.12), the following equation is obtained

$$dF = r^2 \sin\theta \cos\theta (P_0 + \frac{\partial P}{\partial w} R \cos\theta) d\varphi d\theta \tag{5.15}$$

The force of the whole spherule can be obtained by integration:

$$F = \int_0^{2\pi} d\varphi \int_0^{\pi} r^2 \sin\theta \cos\theta \left(P_0 + \frac{\partial P}{\partial w} R \cos\theta \right) d\theta = \frac{4}{3}\pi r^3 \frac{\partial P}{\partial w} \tag{5.16}$$

According to Newton's second law, without gravity, the force on a metal droplet is:

$$F = ma = \frac{4}{3}\pi r^3 \rho a \tag{5.17}$$

where ρ is the density of the metal droplet.

By adopting the scanning electron microscopy (SEM) analysis of the metal deposited on the inner surface of the shield, the components are shown in Table 5.7.

The content of C and O in the table is relatively low and should be impurities, which ought to be ignored. Based on the SEM results and the density of Cu and Cr, the density of the emitted alloy can be calculated according to Eq. (5.18),

$$\rho \approx \frac{\rho_{Cu} \cdot V_{Cu} + \rho_{Cr} \cdot V_{Cr}}{V_{Cu} + V_{Cr}} \approx 7.9 \text{ kg/m}^3 \tag{5.18}$$

Assuming that the droplet's acceleration a is about 10^4 m/s^2, the pressure gradient can be obtained from the above results:

$$\frac{\partial P}{\partial w} = \rho a = 7.9 \times 10^3 \times 10^4 = 7.9 \times 10^7 \text{ pa/m} \tag{5.19}$$

If the pressure gradient along the moving direction of the metal droplet is considered to be constant, and the moving distance l_a during the acceleration of the droplet is about 1.5 mm in the experimental interrupter shown in Figs. 5.16, 5.17 and 5.18,

Table 5.7 Solidification component inside shield detected by SEM

Element	Percentage of mass (%)	Percentage of atomic (%)	Density (kg/m^3)
Cu	54.46	24.56	8.9×10^3
Cr	32.89	26.04	6.9×10^3
C	7.16	24.56	
O	5.48	14.11	

then the pressure from the contact edge to the shield is:

$$P = \frac{\partial P}{\partial w} \cdot l_a = 7.9 \times 10^7 \times 1.5 \times 10^{-3} = 0.118 \text{ MPa} = 1.18 \text{ atm} \quad (5.20)$$

As a matter of fact, there are errors in the above calculation because the pressure gradient cannot be constant everywhere. However, it can be seen from the magnitude of the calculation results that in the vacuum interrupter, there is a large pressure from the center of the contact to the edge, which drives the electrons, ions, and metal droplets to accelerate and diffuse outward. At high current mode, if the density of the plasmas and metal vapors and the number of metal droplets reach a certain level, the arc may break down, which will cause the current breaking to be failed.

In addition, the half-wave arcing time of current at 360 Hz is 1.4 ms. The shortest distance l between the contact edge and the shield is 30 mm in the experimental interrupter. Based on an overall consideration of the former study and Table 5.6, the following assumptions are made: the emitting angle of the metal droplet β is 20°, and the average velocity v_{avg} is 15 m/s. The moving time t from the contact to the glass shield can be calculated by Eq. (5.21):

$$t = \frac{l}{\cos\beta \cdot v_{avg}} = \frac{0.03}{\cos 20\circ \cdot 15} = 2.2 \text{ ms} \quad (5.21)$$

For different frequencies, the calculated results are shown in Fig. 5.20.

For the arc is burning at the first half cycle before CZ, and the dissipation time, which involves particle moving and being solidified after CZ, of the macroscopic particles are longer than the arc time at various frequencies, as displayed in Fig. 5.20. Thus, the breakdown must be impacted by the macroscopic particles which are the source of plasmas and metal vapors post-arc [6].

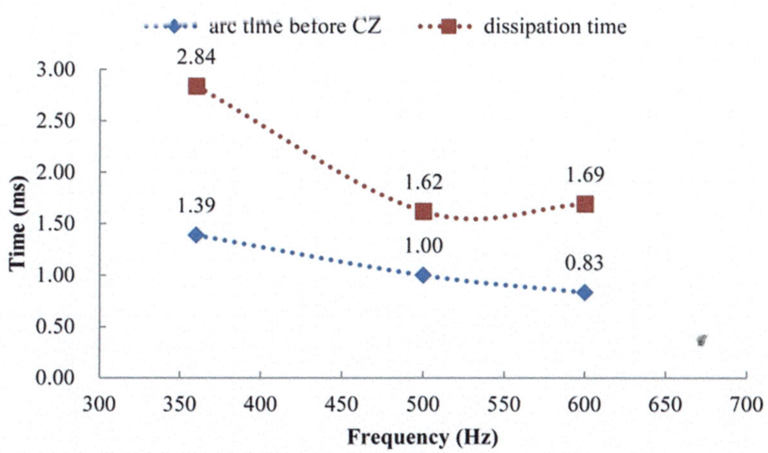

Fig. 5.20 Dissipation time of the macroscopic particles at different frequencies

References

1. Zongqian Shi, Shenli Jia, Lijun Wang, Zheng Wang. Experimental investigation of the relation between vacuum arc voltage and arc appearance [J]. High Voltage Apparatus, 2004, 40(4): 250–252.
2. Lijun Wang, Liuhuo Wang, Shenli Jia, et al. Experimental study of anode activities in high current vacuum arc subjected to axial magnetic fields under different conditions [J]. IEEE Transactions on Plasma Science, 38(7): 1682–1691.
3. Jimei Wang. Vacuum switching technology and application [M]. Beijing: China Machine Press, 2008.
4. Yuan Jiang, Jianwen Wu. Interruption phenomenon in intermediate-frequency vacuum arc [J]. Plasma Science and Technology, 2016, 18(3): 311–318.
5. W. Hartmann, A. Lawall, R. Renz, M. Romheld, N. Wenzel, W. Wietzorek. Cathode spot dynamics and arc structure in a dense axial magnetic-field-stabilized vacuum arc [J]. IEEE Transactions on Plasma Science, 2011, 39(6): 1324–1329.
6. Yuan Jiang, Jianwen Wu, Mingxuan Chen, Wenlei Huo. Analysis of the vacuum arc interruption process in aviation intermediate-frequency power supply systems [J]. Energies, 2017, 10(3):353.

Correction to: Vacuum Circuit Breaker for Aviation Variable Frequency Power System

Correction to:
Y. Jiang and Q. Li, *Vacuum Circuit Breaker for Aviation Variable Frequency Power System*, Power Systems
https://doi.org/10.1007/978-981-33-4781-6

In the original version of the book, the following information has been added: Funding details have been added to the Preface section.

The updated version of the book can be found at
https://doi.org/10.1007/978-981-33-4781-6

Y. Jiang and Q. Li, *Vacuum Circuit Breaker for Aviation Variable Frequency Power System*,
Power Systems, https://doi.org/10.1007/978-981-33-4781-6_6

The manufacturer's authorised representative in the EU is Springer Nature Customer Service Centre GmbH, Europaplatz 3, 69115 Heidelberg, Germany. If you have any concerns regarding our products, please contact ProductSafety@springernature.com

Printed and bound by CPI Group (UK) Ltd, Croydon, CR0 4YY
01/10/2026
02226294-0001